American Wooden Bridges

COMMITTEE ON HISTORY AND HERITAGE
OF AMERICAN CIVIL ENGINEERING

American Society of Civil Engineers
345 East 47th Street, N.Y., N.Y. 10017

1976 | ASCE HISTORICAL PUBLICATION NO. 4 | $12.00

ISBN 0-87262-022-6

INTRODUCTION

Wooden bridges are inimitably American. Their practicality, their individuality, their ruggedness — traditional characteristics associated with our country.

From the first days of colonization, our streams and rivers have been a blessing and a bane. The routes of commerce found them a blessing as long as they were along the water courses, but a bane when needs demanded routes across them. At first they were only traversed at fords, and then ferries were used. Possibly the first bridge crossing in the United States was Coronado's bridge of 1541 which is reported to have been built over the Pecos River, or one of its branches, in New Mexico. Presumably it was a wooden trestle bridge. One tradition indicates that the earliest bridge in the East was one of a similar type built in 1634 over the Neponset River in Massachusetts. However, there is also evidence that a wooden bridge was built in 1611 over the Back River at Frigett Landing near Jamestown, Virginia.

The general evolution of wooden bridges is found in their structural type and the kind of load carried. With respect to load-types, the trend was foot bridges, horse bridges, cart bridges, carriage bridges and then wagon bridges such as those that carried the heavily laden Conestoga's three to five tons. Finally, the early railroads used many wooden trusses, and even today wood trestles are common. Structurally, the sequence seems to have been stringers, beams, trussed beams, trussed-arch, arch, and truss. Wooden trusses were then modified to include iron rods; then iron posts and finally the all-metal truss was introduced in the decade or so preceding the Civil War.

Fire and rot were the main enemies of the wooden bridge. Rot was fought by cladding and paint, but it was the sparking smokestacks and heavy loads of trains that finally forced the major bridges to be built of stone or steel. However, before its demise as the principal bridge material, wood attained the acme of its development in the great "Colossus" of Lewis Wernwag which gracefully arched the Schuylkill near Philadelphia. Built in 1812, this arch-truss spanned 340 feet. (N.B. Burr's wooden bridge over the Susquehanna River at McCall's Ferry, Pennsylvania, which stood from 1815 to 1818, achieved a span of 360 feet, but its relatively short life and isolated location precluded it from receiving the notice of the "Colossus".)

Wooden bridge development did not stop with Wernwag's "Colossus", or Burr's Susquehanna Bridge. It is true that most wooden bridges built during the nineteenth century closely resembled those developed before 1850, but in this century new techniques in laminating and new types of fasteners have enabled wood to once again compete with other engineering materials in bridge construction. For example, near Rapid City, South Dakota stands a wooden (glue-laminated) arch of 155 feet span. Built in the late 1960's, it carries major highway traffic on U.S. 16, and its warm tones harmoniously blend with the forested countryside.

Today, there are about 1000 nineteenth century wooden bridges in over 30 of our states. But they are an ''endangered species''; because their susceptibility to fire and their generally isolated location seems to invite the attention of mindless pyromaniacs and because even Nature, herself, sends rushing torrents to undermine the foundations of poorly maintained structures. Happily, however, many people are concerned with the documentation and preservation of these quiet reminders of our great American engineering heritage. Recently at New England College in Henniker, New Hampshire, a new covered bridge was built to serve both as a stream crossing and a historical reminder. There are many groups actively interested in wooden bridges. Information about them may be obtained from the National Society for the Preservation of Covered Bridges, 63 Fairview Avenue, South Peabody, Mass. 01960.

It is the hope of our Committee that this publication of the American Society of Civil Engineers will not only be of interest to its members, but that it will foster a greater public appreciation of the role of the American wooden bridge in America's development, and thus aid in their preservation as landmarks.

Committee of History and
Heritage of American
Civil Engineering

L. Neal FitzSimons, Chairman
John W. Briscoe
G. Brooks Earnest
Alan L. Prashun
Joseph J. Rady
John R. Wolfs

Contents

CHARLES CONRAD SCHNEIDER

Although trained as a Mechanical Engineer at the Royal School of Technology in his native Saxony, C.C. Schneider found the profession of the Civil Engineer more to his liking and he emigrated to the United States in 1867. Beginning in 1871, when he joined the Michigan Bridge and Construction Company, and ending at his death on January 6, 1916, Mr. Schneider devoted himself to the design and construction of iron and steel structures. Mass production of steel as a construction material was just starting in the 1870's as was Schneider's career, and they both flourished from then on. Although mainly concerned with steel bridges, the challenge of tall buildings (''a vertical bridge truss'') was also met by him. His structural engineering projects were not confined to the United States, but were worldwide; in Mexico, Japan, Egypt and many other countries. When the promising new structural material, reinforced concrete, was in its infancy, Schneider recognized its potential and was among the first to develop engineering standards for its use in bridges and in building frames. So also, he saw the role of wood in the development of bridge structures, hence his care in recounting this role when, as President of the American Society of Civil Engineers in 1905, he gave his presidential address, ''The Evolution of the Practice of American Bridge Building''.

AMERICAN SOCIETY OF CIVIL ENGINEERS.
INSTITUTED 1852.

TRANSACTIONS.

Paper No. 993.

ADDRESS
AT THE ANNUAL CONVENTION AT CLEVELAND, OHIO, JUNE 20TH, 1905.

THE EVOLUTION OF THE PRACTICE OF AMERICAN BRIDGE BUILDING.*

By CHARLES C. SCHNEIDER, PRESIDENT, AM. SOC. C. E.

In the time-honored custom of an address by the President at the Annual Convention of the American Society of Civil Engineers, many of my predecessors have selected topics with which they were most familiar. My life work naturally suggests to me the title of this address.

The magnificent stone bridges and aqueducts built by the Romans bear evidence that the art of bridge building had reached a high degree of perfection among the ancients. Some of the early Roman structures were of gigantic proportions, and have not been excelled by any of those built in modern times. There are no records to show that the ancients had any theoretical knowledge of bridge construction; they built their bridges in accordance with empirical rules developed by experience. These monuments of engineering skill, which have lasted for thousands of years, appear marvelous to the modern engineer, who has at his command, not only theoretical knowledge, but also mechanical appliances and

*Transactions, ASCE, June 1905, Vol. LIV, pp. 213-234

modern tools to assist him, which were not known in ancient times. The progress, therefore, made in the art of constructing bridges of stone in modern times appears insignificant, and is practically confined to theoretical knowledge and improvements in machinery and tools for handling material.

The development of the art of bridge construction is marked to a certain extent by periods coincident either with the introduction of new materials or with the necessities arising for means of transportation. The marvelous progress made in bridge building in the 19th century — the century of engineering, of the manufacture of power, of the railroad and the telegraph — really began with the production of wrought iron in large quantities, and has since kept pace with the development and progress made in the manufacture of those most useful and precious of all metals, wrought iron and steel.

The various types of bridges which have survived and become the standard types of the present day are the results of the evolution of a century, originating with the wooden bridge, the wooden beam being the prototype of the modern plate girder, and the framed wooden truss that of the steel truss. Timber bridges, consisting of hewed logs, supported by piers of piling or of stone, were erected in the earliest times in all countries where timber was abundant, Caesar's bridge across the Rhine was of this character. The wooden bridge consisting of framed trusses is the product of more modern times.

In Europe, wooden truss bridges of small span, patterned after existing roof trusses, have been built for several centuries. In America, however, it was not until the end of the 18th century that the movement began to which the present type of bridges can be traced.

There appear to be no records of any wooden bridges in America before 1785, when unusually gifted men like Palmer, Burr, Wernwag and others commenced to build some very remarkable wooden structures.

The progress made in bridge building in this country was very slow until railroad building commenced, which was in 1829, when the construction of the Baltimore and Ohio Railroad was begun. The first wooden railroad bridge was built on that road in 1830 by

Wernwag, at Monaguay. The development of railroads naturally created a demand for bridges.

The difficulties and obstacles encountered in crossing long and deep valleys were overcome by these pioneer railroad builders by erecting temporary timber trestles in place of expensive embankments or viaducts, to be filled in or rebuilt by permanent structures at a later period. The timber trestle, therefore, is a distinctly American type of construction, and is the prototype of the iron viaduct.

Some of the first high wooden trestles were built in 1840 on the Little Schuylkill and Susquehanna Railroad, now the Catawissa Branch of the Philadelphia and Reading Railroad. They were designed by James F. Smith, and their heights varied from 60 to 130 ft.

The forms of timber trusses of different kinds, arches and combinations of two or more systems, have been very numerous.

A marked step toward bridge designs of the modern truss form was the lattice bridge patented by Towne in 1820, which became the prototype of the early iron lattice bridge.

The next important step in the development of wooden bridges was made in 1840, when Howe patented his truss, which became very popular and the standard for wooden railroad bridges.

In 1844, the Pratt truss was patented, which afterward became the favored type for iron bridges. Many other types of trusses were invented, which have since been discarded.

The earliest wooden bridges were built by expert carpenters. The work was done by contract, very much the same as building work is done at the present day, except that the builder was also the designer. The builder would buy suitable timber or have it sawed to order at conveniently located saw-mills, and any ironwork needed in the construction of the bridge, such as rods, bolts or bars, he would obtain at a local blacksmith shop, and frame and erect the bridge in place, ready for traffic. The same methods were used in building the early iron highway bridges. Each of these builders had his own type of bridge and his own special details. At that time there was generally but little competition, as very few had any knowledge of bridge building, and each one controlled a certain territory.

All the early railroad bridges were wooden structures built by practical carpenters, in most cases employees of the railroad company.

England is considered as the pioneer country of the iron bridge, the first one, consisting of a nearly semicircular cast-iron arch, having been built in 1776-79.

In 1786, Thomas Paine, the well-known author, designed and made a model of a segmental arch. This model was set up at Franklin's house in Philadelphia, whence it was taken to the State House, and, eventually, was sent to Paris and exhibited at the Academy of Sciences. Paine had an experimental cast-iron bridge built in England in 1790, and Rowland Burdon, in 1793 to 1796, built the bridge at Wearmouth, of 240 ft. clear span, after this model, which formed the basis of many cast-iron bridges built thereafter, and became the prototype of the modern steel arch. Paine's device was also the basis of the design of the Market Street Bridge and the first Fairmount Bridge, in Philadelphia, both being wooden arches. The former was completed in 1800, and the latter in 1812.

Up to 1840, there were no iron bridges in this country, except suspension bridges in which iron links were used in the cables and suspenders, the floor-system being of wood.

The first bridge in America consisting of iron throughout was built in 1840 by Earl Trumbull over the Erie Canal, in the Village of Frankfort, N. Y. In the same year Squire Whipple, Hon. M. Am. Soc. C. E., also built his first iron truss bridge.

Probably the first iron railroad bridge was built on the Philadelphia and Reading Railroad at Manayunk by Richard B. Osborne, Chief Engineer, in 1845. It was a double-track through bridge, of 34 ft. clear span, of the Howe truss type, with cast-iron top chord and web braces, the bottom chord and vertical web members being of wrought iron. This bridge was followed by several others of the same type.

The earlier wooden and iron bridges were built very much in the same manner as the ancient Roman bridges, in accordance with empirical rules, by practical men who had no accurate knowledge of the strains produced on the various members of a structure by the exterior forces, but who were men of unusual constructive ability and sound judgment, who had to depend upon their own resources and natural instinct, experimenting with models and profiting by previous failures. Practice always preceded the science, thus the structural systems were invented before their theory was developed.

Until 1847, when Squire Whipple, the modest mathematical instrument maker, who, without precedent or example, evolved the scientific basis of bridge building in America, correct methods of computing the strains in framed structures were not known. A few years later, in 1851, Herman Haupt published a book on the theory of bridge construction.

About 1850, after the building of railroads had advanced, the educated engineer commenced to exert his influence in the art of bridge building, and, from that time forward, steady progress was made.

The period from 1850 to 1860, therefore, may be regarded as an epoch in the history of American bridge building; the time when the bridges designed by Fink and Bollman first came into use, and the earliest iron Whipple and Pratt trusses were built.

When American engineers commenced to build iron bridges, they paid little attention to the then existing European models, but preferred to develop their own systems independently, as they had done previously with wooden bridges, the first iron bridges being imitations of the Towne lattice, and the Howe and Pratt trusses.

THEODORE COOPER

Well known among civil engineers of today for his ''Cooper's Loading'' assumptions for railroad bridges, Theodore Cooper was one of America's leading bridge engineers of the late nineteenth century. Fletcher and Snow called his paper ''American Railroad Bridges'' (1889) ''An authoritative history of [the development of American Timber Bridges]''. It is the earliest one in the Society's transactions.

He was born in Cooper's Plains, New York on January 13, 1839. Graduating from Rensselaer Polytechnic Institute in 1858 at the age of only 19, he entered the field of railroad engineering. His career was interrupted by the Civil War when he entered the Navy as an engineering officer. He remained in service until he resigned in 1872 to join the staff of James Eads who was then building his great St. Louis bridge. Cooper distinguished himself in this work and by 1879 had established himself as a consultant. His clients were many and varied, including industrial firms, construction companies, rapid transit commissions and governmental agencies. In 1907, Cooper retired from active practice and passed away on August 24, 1919.

AMERICAN SOCIETY OF CIVIL ENGINEERS.
INSTITUTED 1852.

TRANSACTIONS.

418.
(Vol. XXI.—July, 1889.)

AMERICAN RAILROAD BRIDGES.*

By THEODORE COOPER, M. Am. Soc. C. E.

The existing and the accepted types of bridges in use today on American railroads being the results of a true evolution, no attempt to present them intelligently would be complete without a brief sketch of the past history of bridges in America.

The rapid development of the new world and the enormous number of bridges that has been built within the limits of the nineteenth century, have furnished us with a wide experience, from which we have been able to select the good and reject much that was bad or undesirable.

The pioneer life, not only of the earlier settlers, but of each generation to the present day, has developed to a high degree the energies, ingenuity and self-reliance of the American people. These pioneers were compelled to be men of all trades. Their limited resources and the lack of time or opportunity to seek for past precedents, impelled them to solve each problem anew. They, "thought with vigor and were not fettered with the trammels of science, before they were capable of exerting their mental faculties to advantage," as Sir Joseph Banks wrote to *Thomas Paine* in 1788.

*Transactions, ASCE, July 1889, Vol. XXI, pp. 1-58

Having no educated "lines of least resistance," they were better able to solve the many problems before them by new and novel methods.

The bridging of small streams was a part of the pioneers' labor. The crossing of the larger rivers developed specially gifted men, like *Timothy Palmer, Theodore Burr, Lewis Wernwag,* and others less well known, who built timber bridges that are looked upon as wonderful structures, even to the present day. The records of the early bridges of America are very incomplete, but enough remains to show what admirable work these early bridge builders could do.

1. Wooden Bridges.

The earliest bridges, where single timbers were not sufficient to stretch from bank to bank, were short spans supported on piles, or, where these could not be used, on timber cribs filled with stone. Where the conditions would not allow of structures of this character, arch spans were usually adopted.

In 1660, "The Great Bridge," as it was then called, was built across Charles River, between Old Cambridge and Brighton. It was a pile bridge.

In 1761 *Samuel Sewall* planned and built a bridge over York River, Maine, 270 feet long, supported on thirteen piers. Rebuilt in 1793.

In 1786 Mr. Sewall built a bridge over the Charles River, at Boston, 1,503 feet long, supported on seventy-five piers. A year or so later bridges on the same plan were built at Malden and Beverly, Mass.

In 1792 *Colonel William P. Riddle* built the Amoskeag Bridge at Manchester, N. H. It was 556 feet long, and supported on five piers and two abutments. It was commenced on the 3rd of August, "at which time the timber was growing, and the rocks dispersed in the river," and completed on September 29th.

Between 1785-92 Colonel *Enoch Hale* built over the Connecticut River, at Bellows Falls, a bridge 368 feet long, in two spans, taking advantage of a rock in the middle of the river for his center pier. The West Boston Bridge over the Charles River, 3,583 feet long, and supported on one hundred and eighty pile bents, was finished in 1793.

Near the end of the eighteenth century a bridge was built over Cayuga Lake, N. Y. It was a pile bridge in 25-foot spans, one mile in length.

In 1795 a bridge was built over the Mohawk River, 960 feet long, supported on thirteen piers.

In 1792 *Timothy Palmer* built the Essex-Merrimack Bridge over the Merrimack River at Deer Island, about 3 miles above Newburyport, Mass. It consists of two bridges resting on Deer Island in the midst of the river (PLATE I). "An arch of 160 feet span and 40 feet above the level of high-water connects this island with the mainland on one side; the channel on the other side is wider, but the center arch is but 113 feet." That part of the bridge on the Newbury side, the 160-foot span, was removed in 1810 and replaced by a chain suspension bridge. The part on the Salisbury side remained until 1883. The chain bridge was built by *John Templeman,* of the District of Columbia. "It was the first chain bridge built in New England." Its span is 244 feet between bearings on towers; the towers are timber frames covered with boards and shingles. February 6th, 1827, one of the chains broke under a heavily loaded wagon drawn by four oxen and one horse. The bridge was again rebuilt, and is still in use. The bridge, as it now exists, consists of two independent roadways, each 15 feet 6 inches wide. Each roadway is supported on two sets of chain cables, each set being composed of three chains. These chains seem to have been repaired in many places with different sized links. The chains generally are formed of links about 2 feet long, made of 1-inch square irons. For about 6 feet over the bearings on the towers, each chain is spliced or replaced by three smaller chains with links about 1 foot long, and of about ½-inch square iron. The floor is hung from these chains every 7 feet by suspenders, formed indifferently of bars 1 inch square, straps 2 x ½ inch, or pieces of chains (PLATES II AND III).

In 1793 *Timothy Palmer* built another bridge over the same river at Andover. Rebuilt in 1803.

In 1794 he built the Piscatauqua Bridge, 7 miles above Portsmouth, N. H. It was 2,362 feet long. The greater part consisted of pile work.

"But that part which engages the attention of travelers, is an arc nearly in the center of the river, uniting two islands over water 46 feet deep. This stupendous arc of 244 feet chord is allowed to be a masterly piece of architecture, planned and built by the ingenious Timothy Palmer, of Newburyport, Mass."

"The chord of this arch is 244 feet 6 inches. The versine of the arch was 27 feet and 4 inches, and depth on the frame-work of the arch

18 feet and 3 inches. There are three concentric ribs, the middle one carrying the floor of the bridge; they were selected from crooked timbers, so that the fiber might run nearly in the direction of the curves, and are connected together by pieces of hard and imcompressible wood. with wedges driven between, the ribs being mortised to receive them; thus the ribs are kept at a regular and parallel distance from each other. Each rib is formed of two pieces laid side by side about 15 feet in length; they are all disposed in such a manner as to break joints, the end of one timber coming in the middle of the length of the other which is near it; their ends all abut with a square joint against each other and are neither scarfed nor mortised, the two pieces of timber being held together by transverse dovetail keys and joints; all the timbers are admirably jointed and freely exposed to the action of the air; any piece may also be removed in case of its requiring separation without injury to the rest of the structure."

The bridge was 38 feet wide and had three arched trusses. Another description states that the second rib carrying the floor of the bridge was of a larger radius than the lower to facilitate the traveling; the upper rib served for a railing.

In 1794 the bridge at Haverhill was built by *Timothy Palmer*, "contemporaneous with the first stage coach and the first newspaper." It consisted of three arches of 180 feet; each is supported on three handsome piers 40 feet square; it had as many defensive piers or sterlings extending 50 feet above and a draw 30 feet over the channel.

In 1796 he built a bridge over the Potomac River, at Georgetown.

In 1796 *Rufus Graves* built a bridge over the Connecticut River, at Hanover, N. H., consisting of a single arch of 236 feet. He patterned his bridge after the Piscatauqua Bridge built by Palmer. The roadway followed the line of the arch and was some twenty feet higher at the center than at the abutments. The bridge was formed of the largest selected white pine, hewed to 18 inches square, some of them 60 feet long. This bridge fell in 1804, without warning, and by its own weight; its destruction being hastened by the undermining of one of the abutments, through deficient waterway. The builder of this bridge studied divinity, graduated at Dartmouth College, became a merchant, then a bridge builder, then an officer of the United States Army, and finally a physician.

In 1795 a bridge was erected at Holt's Rock, between Newbury and

Haverhill. It was 1,000 feet in length and consisted of four arches and one draw-span. It was carried away by the ice in 1818.

In 1796 a bridge was built between Harlem and Morrisania, over the Harlem River.

About the end of the last century a bridge was built at Windsor, Vt., with two spans of 144 feet, and one at Romley, Mass., consisting of 8 arches of a total length of 870 feet; one at Howland's Ferry, R. I., 900 feet long, with a sliding draw, supported on 42 pile bents, and the Weybasset bridge at Providence, R. I., "160 feet long, supported by two wooden trestles and two stone pillars."

There were also two bridges over the Lehigh River, one at Bethlehem and one at Easton.

These are the records as I have been able to gather them, of bridges of any importance built in the eighteenth century.

From 1804 to 1806 "The Permanent Bridge" over the Schuylkill River at Philadelphia was built. It consisted of two arches of 150 feet clear and one of 195 feet clear.

"The plan was furnished by *Mr. Timothy Palmer,* of Newburyport, Mass., a self-taught architect. He brought with him Mr. Carr as his second and four other workmen from New England. They at once evinced superior intelligence and adroitness in a business which was found to be a peculiar art, acquired by habits not promptly gained by even good workmen in other branches of framing in wood." "The frame is a masterly piece of workmanship, combining in its principles that of king post and braces or trusses, with those of a stone arch."

There were three truss and arch frames. The timber was of the best white pine. Width of the bridge, 42 feet. The bridge was covered and closed in from the weather. Mr. Palmer stated that from his experience, wooden bridges uncovered would become unsafe in ten to twelve years.

"I am an advocate for weather boarding and roofing, although there are some who say it argues much against my own interests; notwithstanding I am determined to give my opinion as appears to be right. It is sincerely my opinion that the Schuylkill Bridge will last thirty and perhaps forty years if well covered. You will excuse me in saying, that I think it would be sporting with property, to suffer this beautiful piece of architecture (as you are sometimes pleased to call it), which has been built at so great expense and danger, to fall into ruin in ten or twelve years."

PLATE IV is copied from an old engraving. The superstructure was entirely renewed on a more modern plan of a wooden arch bridge in 1850, and widened for also carrying a railroad track. The foundations for this bridge were very difficult of execution, owing to the lack of experience at that day in work of this kind. Depth of water at east pier 24 feet, and at west pier 41 feet, rock bottom with slight covering. The foundations were laid in coffer-dams. *Mr. William Weston,* an eminent English engineer, who was in Philadelphia at that time, furnished the plan for the coffer-dam; but "declared that he should hesitate to risk his professional reputation on the event." Afterwards, in a congratulatory letter to the President, he says:

"I most sincerely rejoice at the final success that has crowned your persevering efforts in the erection of the west pier; it will affort you matter of well founded triumph when I tell you that you have accomplished an undertaking unrivaled by anything of the kind that Europe can boast of." "I have never, in the course of my experience or reading, heard of a pier founded in such a depth of water on an irregular rock affording little or no support to the piles."

This pier, however, was not carried to the rock, for the weakness and leakage of the coffer-dam became so threatening, the masonry was started three or four feet above the rock.

In 1805 *Timothy Palmer* completed the bridge over the Delaware River at Easton, Pa. It consists of three spans, 163 feet clear. This bridge, after eighty-four years' service, is still in use and about five-sixths of the original timber is in good condition. It has always been covered in from the weather. (See PLATE V.)

In 1804, *Theodore Burr* built the Waterford Bridge over the Hudson River, consisting of four arch spans, one 154 feet, one 161 feet, one 176 feet, and the other 180 feet, clear spans. The timber is hewn yellow pine. This bridge was not covered until 1814, but is still in use, and in reasonably good condition. (See PLATE VI.) The footings of some of the arches required renewal a couple of years ago. The drawings of the Easton and Waterford bridges are made from sketches taken from the existing structures. Burr also built another bridge over the Hudson at Fort Miller.

A bridge over the Connecticut River at Springfield was built about this same date by *Mr. Walcott.*

From 1804-06, Theodore Burr built the bridge over the Delaware River at Trenton (PLATE VII) consisting of five arch spans, two of 203 feet, one of 198 feet, one of 186 feet, and one of 161 feet in the clear. Each span had five arched ribs, formed of white pine plank, from thirty-five to fifty feet in length and 4 inches thick, repeated one over the other, breaking joints, until they formed a depth of 32 inches. The lower chord was composed of two sticks 6½ x 13½ inches. The roadway was suspended from the arch ribs by vertical chains. The arch was counter-braced by diagonal braces, formed of two sticks 6 x 10 inches spiked to the lower chord and secured to the arch above by iron straps. Outside of the exterior ribs wing arches 50 feet long, splayed out so as to widen the bearing on the piers and abutments. The bridge had no other wind bracing. This bridge is erroneously credited to Lewis Wernwag. The arch footings required renewal in 1832, owing to decay of the timber. In 1848 the bridge was remodeled by removing the wing arches and adding a new and stronger arch rib on the south side, and at the same time strengthening the adjacent old arch rib, by increasing its depth to 4 feet, to carry a railroad track. In 1869 it was again strengthened. It was finally removed in 1875, and replaced by an iron structure.

This bridge may, therefore, be considered as a connecting link between the early wooden long-span bridges for highway traffic and those for railroad purposes.

In 1808, *Theodore Burr* built the bridge shown on PLATE VIII, at Schenectady, N. Y. This was the second bridge built at that place by Burr, the first falling before or soon after completion. The traditions in regard to the first bridge are not very reliable. The contract for the one shown, however, states that it is to be "exactly on the site of the bridge built by the said Theodore at the same place," and apparently with different spans, as new piers are provided for, by the contract. There may, therefore, be some truth in the tradition, that the first bridge was to consist of two spans of 450 feet; that one span was completed and the other partially, when it fell or was swept away by floods.

The bridge shown is a curiosity in bridge construction. It is a suspension bridge of wood. The curved ribs are formed of eight 4 x 14-inch planks spiked or bolted together. The planks are cut to long bevels at the ends for splicing. The ribs are three in number, spaced about 13 feet apart. The timber was white pine. The plan shown is copied from Burr's original drawing attached to the contract.

The bridge was used for twenty years in this condition. It then became necessary to build additional piers under the middle of each span to stop the excessive sagging of the bridge. In this condition it stood till 1873, when it was replaced by an iron bridge.

From 1812-16, *Theodore Burr* built the bridge at Harrisburgh, Pa., over the Susquehanna River, consisting of twelve spans of about 210 feet. The half of this bridge which is south of the Island still remains in use.

In 1812 *Lewis Wernwag* built his "Colossus" bridge over the Schuylkill at Fairmount, Philadelphia. (See PLATE IX.) It had a clear span of 340 feet 3¾ inches. The eastern abutment was founded on the rock, and the western one on piles driven to the rock. This bridge was destroyed by fire, September 1st, 1838.

In 1814, he built the New Hope Bridge, PLATE IX, over the Delaware River, consisting of six arch spans of 175 feet; versed sine of arch, 13 feet. Each span consisted of three arch ribs, formed of four sticks 6 x 15 inches, dressed to the curve and confined by iron clamps and bolts.

Wernwag's practice was to saw all his timbers through the heart, to detect unsound wood and permit good seasoning of the timbers. He used no timbers of a greater thickness than 6 inches, and separated all the sticks of his arches by cast washers, to allow free circulation of the air.

In 1810 he built a wooden cantilever bridge over the Nashammony River, Pa. He called it his "Economy" bridge, and claimed that it could be used to advantage up to 150 feet spans.

From this date to 1836 he built many bridges. In 1830 he built his first railroad bridge at Manoguay, on the Baltimore and Ohio Railroad.

His last bridge was the railroad bridge over the Canal and Potomac River, at Harper's Ferry, in 1836.

The previously mentioned bridges with spans of any magnitude appear to have been arch bridges.

The first marked step toward bridges of the modern truss form was the wooden lattice bridge, patented by Ithiel Town, January, 1820. PLATE X.

The horizontal members or chords of this bridge were compsed of two or more parallel sticks, spaced so that the diagonal web members or lattice work would pass between them. The timbers were usually 2 to 4 inches in thickness, and 10 to 12 inches wide. The web members were placed closer together for the longer spans. The chords were also made in two sets for the longer spans, one placed above the other.

The web members at every crossing and at their junction with the chord pieces were fastened with wooden treenails.

The timbers being of one size and of reasonable dimensions, were readily obtainable. The absence of all bolts, straps or rods of iron, and the simplicity of the mechanical operations required to connect together the parts of this bridge, made it a cheap and popular structure.

A great number of bridges of this class, up to spans of 220 feet, were built throughout the United States, both for highway and railroad purposes. There are many of them still in existence.

These bridges were made of uniform section throughout. In order to get sufficient support at the ends, they were usually extended over the abutment a distance about equal to their depth, and were made continuous over all the piers.

From the thinness of the web system the trusses were apt to warp, and as they aged they became very flexible, owing to the want of rigidity of the treenail connections.

Mr. Town claimed, at least in his later pamphlets (1831), that these trusses could be made of wood, wrought or cast-iron.

The next step toward simplicity and concentration of parts appears to have been the truss known as the Long truss, patented by Brevet-Lieutenant-*Colonel Long* of the United States Engineers, in March, 1830, and November, 1839.

This also was a form of truss in which iron did not enter as a necessary part, the connections being made by framing the parts together, or by use of wooden keys or treenails. Many bridges were built of this form, but it never became widely popular.

Both the Lattice and the Long bridges were combined with the arch in many cases, especially for the longer spans.

No further important advance in the styles of the trusses was made until 1840, when *William Howe* patented the truss which was the basis for the present Howe truss bridge. PLATE XI.

This form of truss grew rapidly into favor, from its simplicity of construction, perfection of detail and satisfactory action under service. For some years it has been the standard form of wooden bridge in use upon our railroads.

The chords and braces are made of timber, and the vertical web members (tension) of round iron, with screw ends, usually upset to give a stronger section through the screw threads than in the body of the rods. The chords are made of uniform section throughout, and both the top and bottom chords of the same number of timbers, and timbers of the same width and spacing. The sticks of the lower chord (for through bridges) are made deeper than those of the top chord, on account of the loss of section from splicing, the lower strains used in tension and to resist the bending strains due to the cross floor timbers, which rest directly on the lower chord. The chord sticks are spaced two or more inches apart, to allow the passage of the vertical rods without cutting away any of the timber, to give room for the packing and splice blocks, and to admit free circulation of air between the timbers. The number and length of the chord timbers are arranged so that no more than one stick need be spliced in any one panel of the truss. The chord sticks may be spliced either with oak or iron splices.

The braces and counterbraces are square-ended, and rest upon the inclined faces of the angle blocks. They are held from displacement by lugs or dowels. The angle blocks were formerly made of oak, but now are made of cast-iron. They are provided with square faced projections, which are let into the chord timbers, to take up the horizontal shear of the braces by direct fiber compression.

The "gib plates," as the plates on which the nuts of the web rods bear are called, are usually made of thick, flat plates of wrought-iron.

The space between the chord timbers below the angle blocks contains cast tube blocks, made a little short to allow for the shrinkage of the timber.

The vertical shear of the braces is transferred through the angle blocks and these tube blocks to the gib plates and web rods, thus avoiding any crushing action across the fiber of the wooden chord pieces.

The timber in a properly designed Howe truss is only strained longitudinally with the fiber.

The very wide experience in this class of bridges has enabled the American bridge builders to so proportion them in all their details that the best results can be obtained from the material at a moderate cost. But for the two objections, their liability to destruction from fire or from decay, no better railroad bridge up to 150 feet spans could be desired. For most of our country the cost of good timber makes iron bridges the cheaper, at first cost, for spans over 150 feet.

Very few Howe truss bridges are now being built upon the older and financially better roads, all wooden bridges being replaced with iron structures.

In April, 1844, *Thomas W.* and *Caleb Pratt* patented the truss known as the Pratt truss. It differed from the Howe in making the diagonal members of the web system of iron (tension) and the vertical members (compression) of wood, the reverse of the Howe principle. While many bridges were built upon this form, it never succeeded in attaining an equal popularity with the Howe; as it required a greater quantity of the more expensive material, wrought-iron, and was not so well suited to the joint use of the two materials, wood and iron.

It became, however, the favored form, afterwards adopted, for iron bridges, and is therefore one of the steps in the development of American bridges.

In addition to the general forms of structures heretofore mentioned, the variations of these broad types and other special forms were numerous. Each individual builder had his own favorite at some time. It would be impossible and useless to take up those numerous individual cases, though they undoubtedly served a good purpose in the general evolution.

The development of wooden bridges was entirely empirical. They were generally of uniform section, and their proportions determined either from the result of previous failures or from the study of models. The advancement was, however, toward simplicity of form and construction, and greater perfection of the details of the connection of the parts.

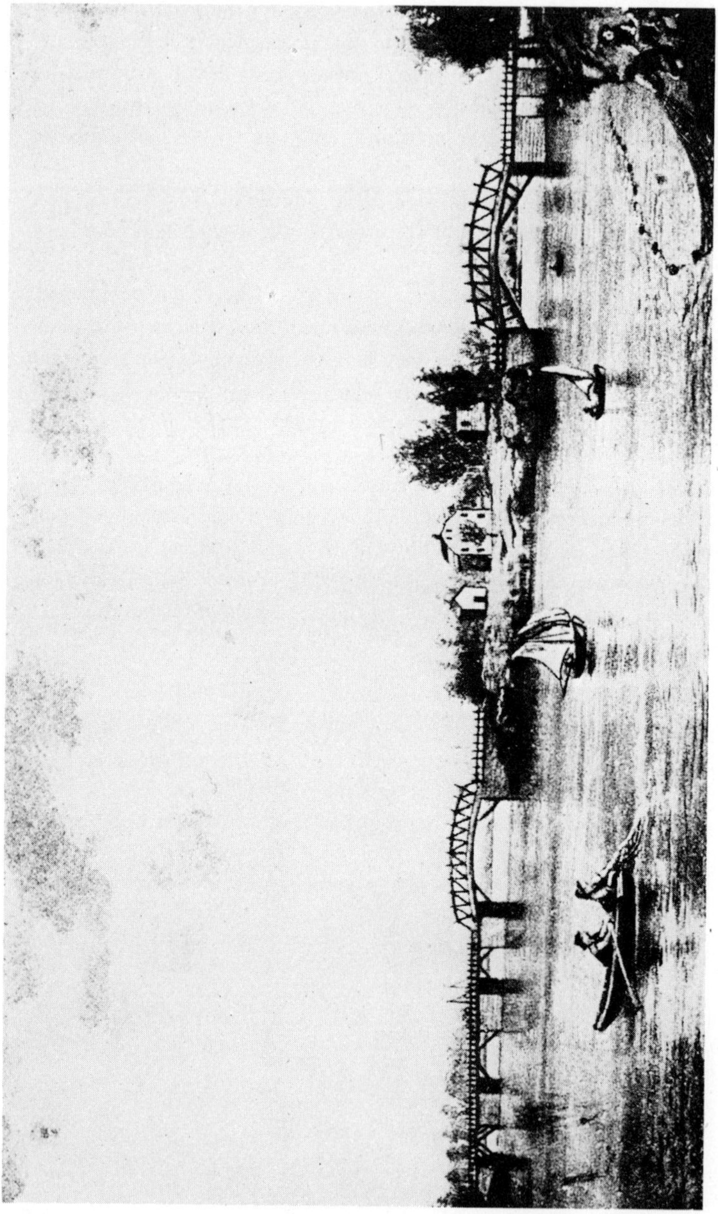

Plate I. ESSEX-MERRIMACK BRIDGE. 1792. Vol. XXI, p. 4.

Plate II. ESSEX-MERRIMACK BRIDGE. 1810. Vol. XXI, p. 4.

PLATE III. ESSEX-MERRIMACK BRIDGE. 1810. Vol. XXI, p. 4.

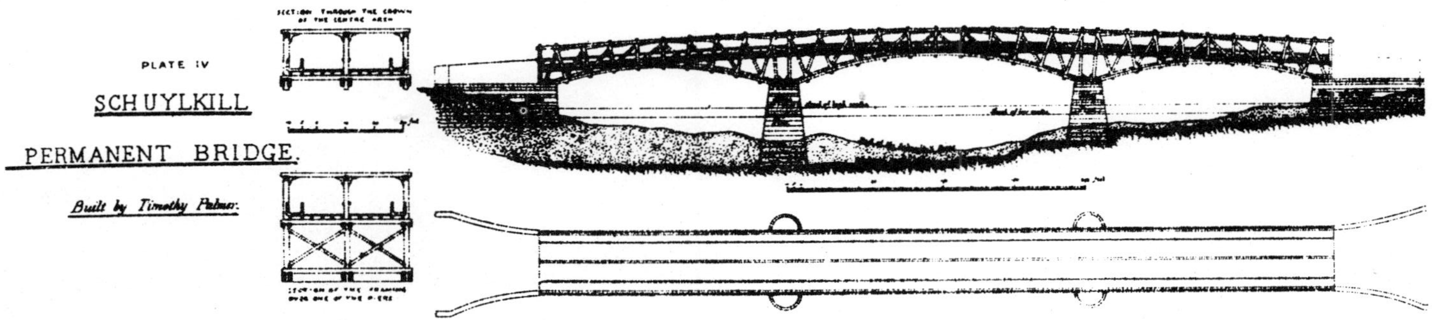

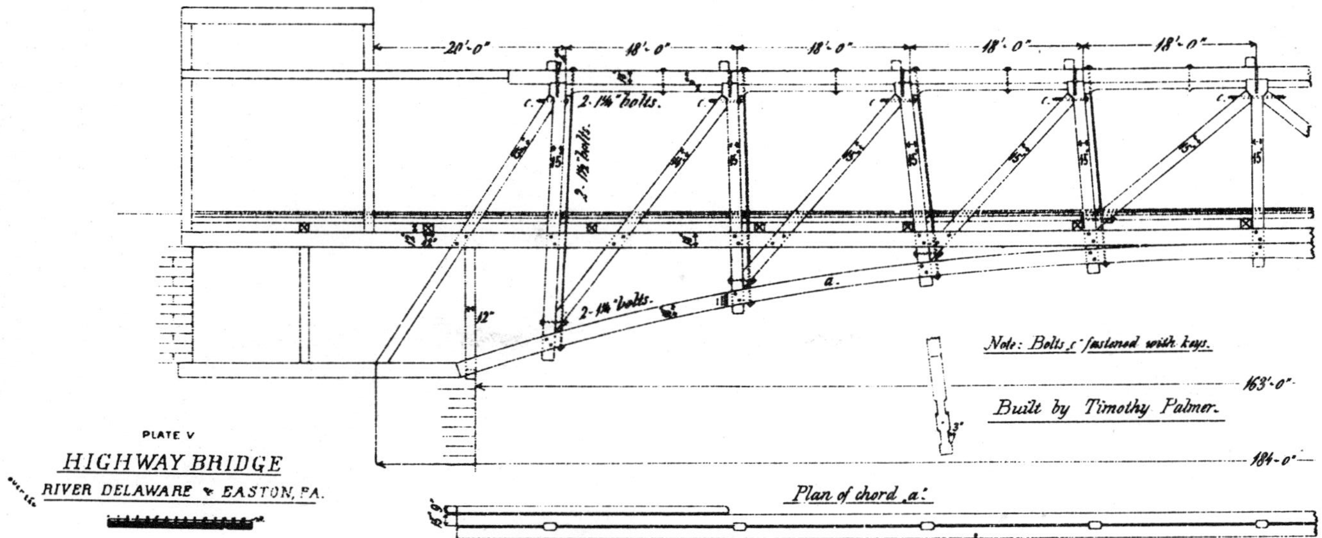

TIMOTHY PALMER'S BRIDGE AT EASTON, PA., 1805

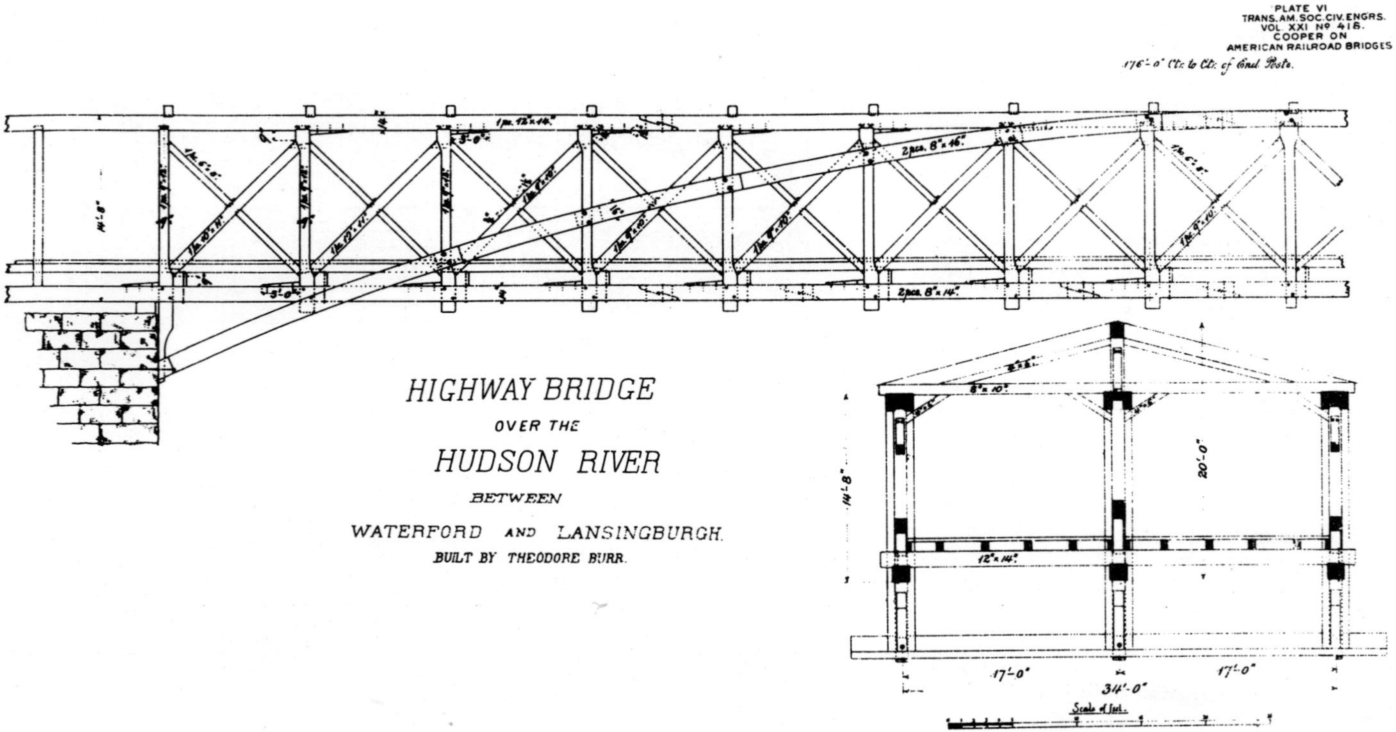

WATERFORD BRIDGE OVER HUDSON, 1804

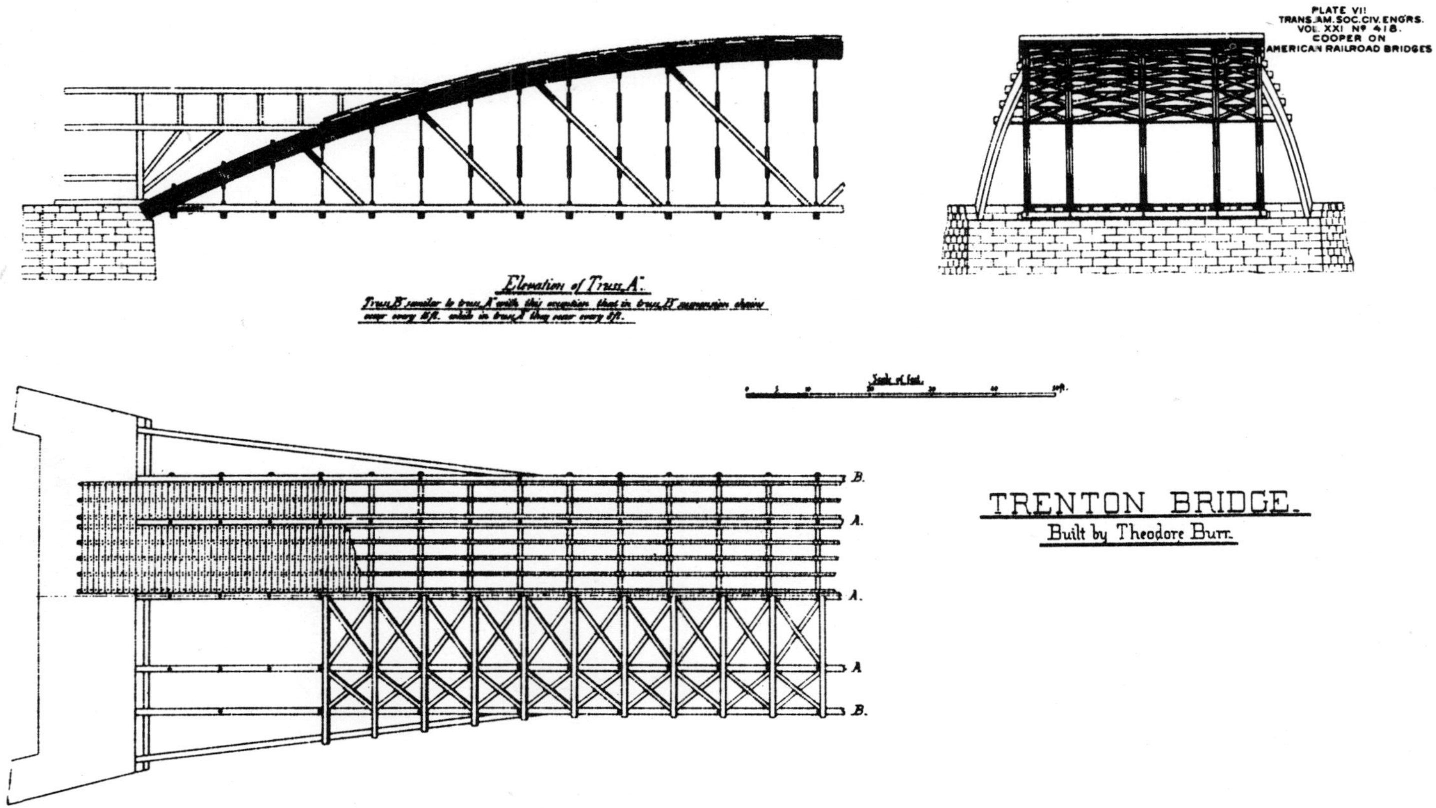

DELAWARE RIVER BRIDGE AT TRENTON, 1804-06

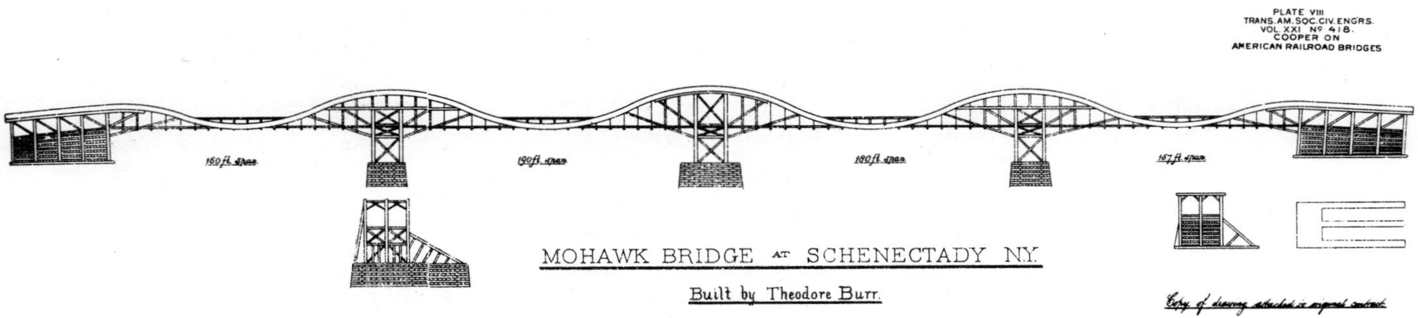

BURR'S SCHENECTADY, N.Y. BRIDGE, 1808

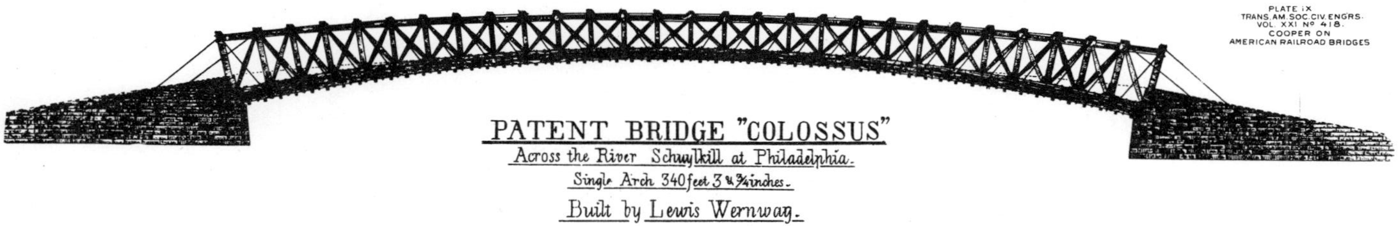

WERNWAG'S "COLOSSUS" BRIDGE AT UPPER FERRY, PA., 1812

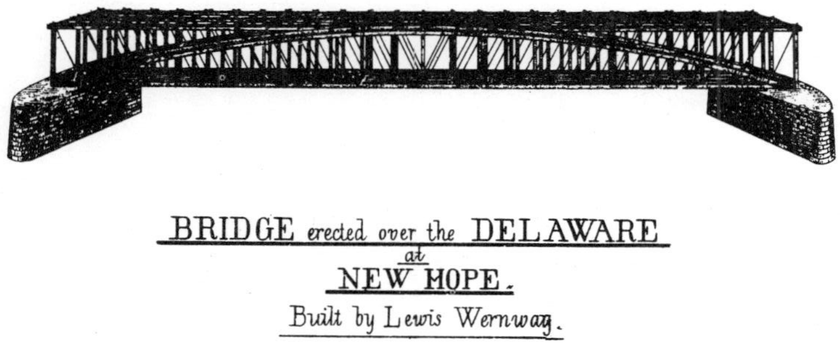

WERNWAG'S DELAWARE RIVER BRIDGE AT NEW HOPE, PA., 1813-14

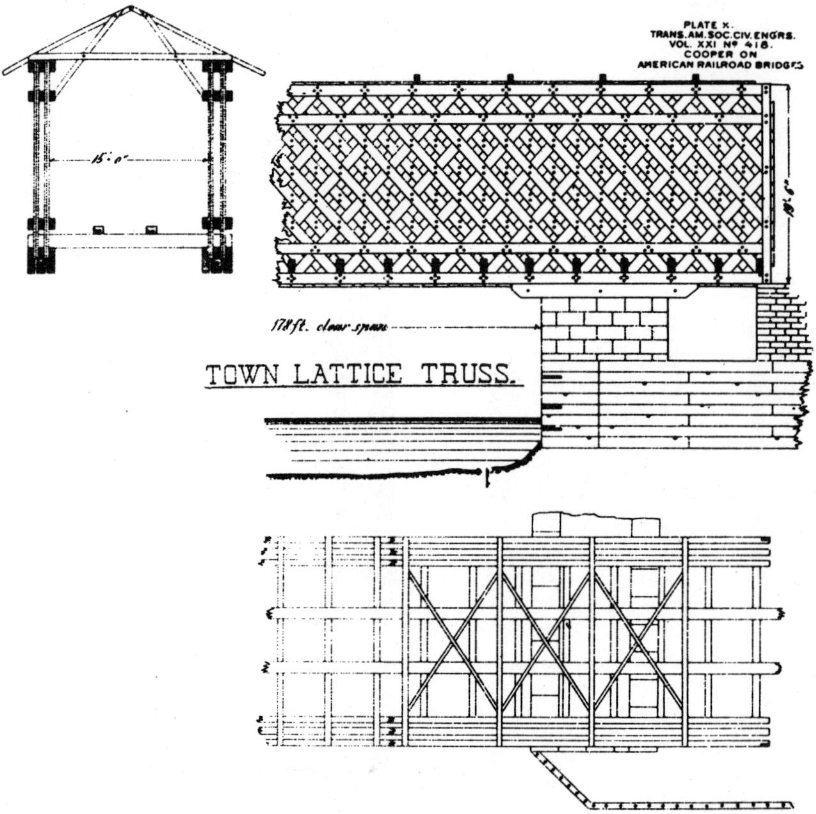

TOWN'S WOODEN LATTICE BRIDGE PATENTED 1820

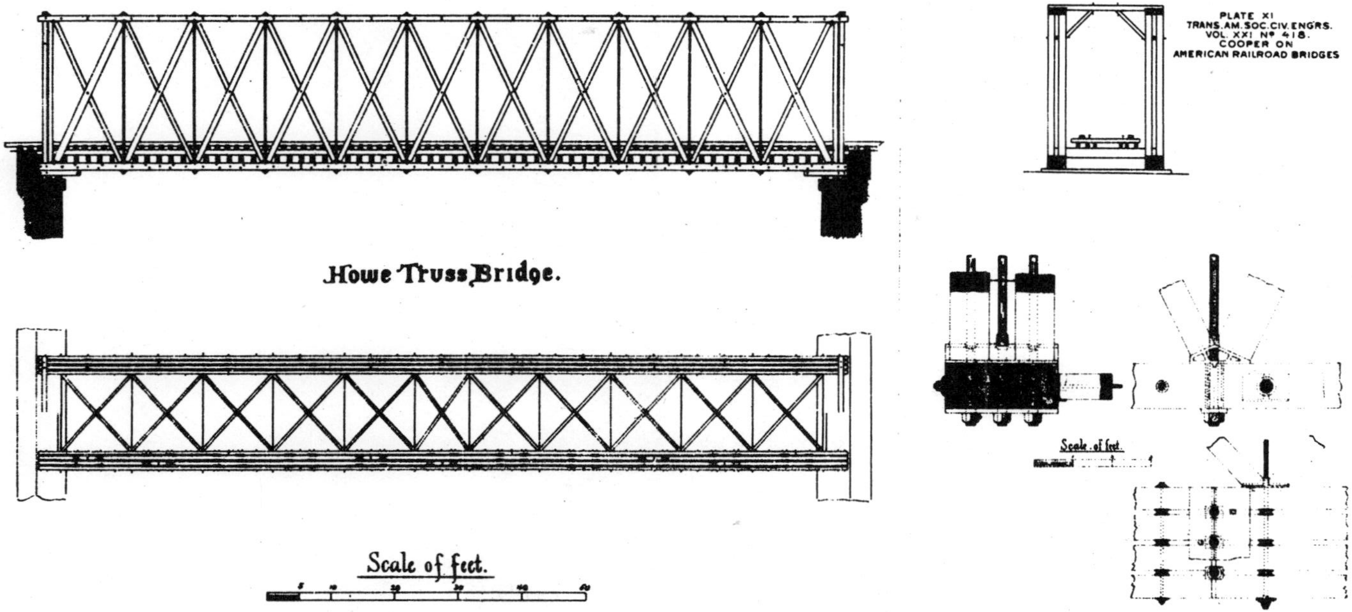

HOWE TRUSS BRIDGE PATENTED 1840

ROBERT FLETCHER & JOHATHAN P. SNOW

The writing team who authored, the classic "History of the Development of Wooden Bridges" were perfectly matched for the task. Mr. Fletcher, an engineering educator, contributed his scholarship; Mr. Snow, an engineering practitioner, contributed his practical experience.

Curiously, although Robert Fletcher was born only a year earlier than Jonathan Snow (August 23, 1847 versus November 19, 1848), he was Snow's professor when he graduated from Dartmouth in 1875. The reason for this unique situation was that the young Mr. Fletcher, after a brief career as an army officer (USMA '68), was selected by President Smith of Dartmouth to organize the Thayer School of Civil Engineering. Smith had been persuaded to select Fletcher by none other than Sylvanus Thayer. This was in 1870 when Fletcher was only 24 years old. He completely justified both President Smith's and General Thayer's confidence and served on the Dartmouth faculty for 47 years.

J.P. Snow was 27 years old when he received his degree of Civil Engineer. The railroads of New England were Snow's arena of professional activity, first on company staff and later as a consulting engineer. His specialty was railroad bridges, particularly wooden bridges; he was their staunch advocate especially over iron bridges. When steel became a widespread bridge material, he was among the few American engineers who favored the rigid joint over the "pin" joint. Mr. Snow died on September 4, 1933 just a few weeks after the reprint of the "History of the Development of Wooden Bridges" was published.

Mr. Fletcher outlived his former student, but on January 7, 1936, he passed away in his beloved Hannover, New Hampshire which had been his home for 65 years.

AMERICAN SOCIETY OF CIVIL ENGINEERS

Founded November 5, 1852

TRANSACTIONS

Paper No. 1864

A HISTORY OF THE DEVELOPMENT OF WOODEN BRIDGES*

By Robert Fletcher,[1] and J. P. Snow,[2] Members Am. Soc. C. E.

With Discussion by Messrs. Henry B. Seaman, Jasper O. Draffin, E. K. Morse, John W. Storrs, C. J. Hogue, William A. Oliver, Philip G. Laurson, Richard S. Kirby, Wells N. Thompson, Benjamin Wilder Guppy, Ivan E. Houk, William H. Baker, Jr., Oren Reed, R. P. Davis and L. L. Jemison, Peter T. Landsem, J. K. Finch, A. A. Eremin, Aksel Anderson, William G. Atwood, H. Fugl Meyer, W. E. Howland, Harry K. Ellis, Nelson J. Bell, and Robert Fletcher and J. P. Snow.

Synopsis

This story of wooden bridges is written from the engineer's viewpoint. Other writers have presented the æsthetic, non-technical, and sentimental aspect of the subject. Beginning with the dawn of history, the story ends with the evolution of types of timber bridges accomplished only within the two centuries just past; and this development has culminated in the United States, giving to this country the latest types designed in conformity with correct principles and built at minimum cost. In the treatment of the subject three periods may be recognized.

The first period covers what it is now possible to learn concerning wooden bridges constructed by so-called "ancient" people and includes the indefinite period indicated by the historian's phrase, "the fall of the Roman Empire." There seems to be reason for the opinion that types of primitive bridges have survived many centuries in the "unchanging East," and only recently have come distinctly to the knowledge of the Western civilized world. Descriptions of such types are given.

*Proceedings, ASCE, November, 1932, Vol. LVIII, pp. 1455-1498

[1] Director Emeritus, Thayer School of Civ. Eng., Dartmouth Coll., Hanover, N. H.

[2] Cons. Engr., Boston, Mass. Mr. Snow died September 4, 1933.

The second period covers the development in Europe from the so-called "Dark Ages" to about the Sixteenth Century, when scientific knowledge began to influence the minds of mankind—the period of the Renaissance. Then, beginning about the middle of the Eighteenth Century, a variety of types appeared, showing much ingenuity in design and skill in construction. It was a time when both rulers and people gave more attention to public works of all descriptions.

Finally, the modern period deals at greater length with the various stages of development through experimental types of American timber bridges to types now accepted as standard. Such are principally those which make use of lattice, Howe, and Pratt forms of truss, with or without a timber arch in combination.

Presumably the end of the period would be when steel and concrete have displaced timber in bridge construction; but it will appear that, where timber is plentiful and considerations of economy and rapid accomplishment prevail, wooden bridges still have "a reason to be." A review of the dominant facts established is given in the "Summary."

Introduction

Extensive inquiries made in twenty-four (only one-half) of the United States (Eastern, Middle, Southern, and Central) have revealed the surprising fact that, as late as 1930, there were between 450 and 500 covered bridges in use, a goodly number of which are likely to serve for some decades to come. A quarto volume, published in 1931, gives 135 illustrations of such bridges.[3] The ultimate settlement of a large area of undeveloped country, such as Alaska, South America, and parts of the Eastern Hemisphere, where timber is abundant, will make demands for bridges that can be built with timber, quickly and cheaply, and that will be durable enough for the pioneer period.

Originally, much of the area of the United States was heavily forested, especially the Eastern part. Naturally the first settlers used the timber at hand for bridges and other construction. The shipyards along the New England Coast, where so many fine ships were built, were effective training schools for heavy timber framing; and the influence of that training still remains among the carpenters of the region. Those conditions led to the building of many bridges of note during the period of development and trade after the Revolutionary War. The spirit of emulation and invention which is characteristic of the American people resulted in a flood of patents for bridge-truss types, a few of which have led to good designs which have been widely approved and adopted. The full text of the investigation upon which this paper was based has been placed on file with Engineering Societies Library, in New York, N. Y., to serve as a possible aid to any subsequent research on this point.

[3] "Covered Bridges in America," by Rosalie Wells, with foreword by Charles S. Whitney, M. Am. Soc. C. E., New York, William Edwin Rudge, 1931.

The Ancient Background

Undoubtedly the prehistoric man built his bridges with the adaptable materials within his environment. Where trees abounded, a log fallen or felled so as to bridge a stream, would be a suggestion to the most primitive intelligence. Hence, the simple beam bridge, under favoring conditions, would be a first resort and, from prehistoric time to the present, may be regarded as a world-wide type.

Again, if the tale[4] of an early explorer in South America may be believed, even the monkey may have taught the prehistoric man. In a tropical forest he may have seen a chain of monkeys, their bodies locked together by means of their long arms and long tails, hanging from the branch of a tall tree and swaying until the lowermost grabbed a branch on the other side of a stream. Then the crossing of the group on this living bridge, the shifting of the point of attachment on that side, the reverse swing of the chain until the rearmost monkey landed, and, finally, the breaking of the chain would be quite an effective demonstration, if and when witnessed.

In the sub-tropical parts of Central Asia, claimed to be the cradle of the human race, bountiful Nature provided ample material for constructing suspension bridges. There is the rattan, or climbing palm, a species of the *genus calamus,* remarkable for the length of its stems growing out from a slender trunk.[5] These stems are very tough, having hooks by which they climb over bushes and trees; sometimes they attain a length of 300 to 800 ft., or more.

Another adaptable material is found in the liana, defined as any perennial plant having woody stems, which are characteristic of certain geographical regions, as the forests of the Amazon, the jungles of India, etc.[6] In the temperate zone the grapevine is an analogous example. The grass rope is a species of bamboo that belongs to the great grass family.

Rope bridges may be classified into five types: (1) Tight rope; (2) single rope and dragline; (3) toboggan; (4) roadway; and (5) tubular type. The tight-rope type, consists of a single rope to serve as a footwalk and two side ropes for hand-rails. This type has been mentioned by the English traveler, E. Kingdon-Ward, to illustrate his article[7] on "Bridge Builders of the Jungle." It is a remarkable fact that this same type of bridge was found by an American explorer in the Cameroon—the French mandate in Equatorial West Africa.[8] It was made safer, however, by a woven net of vines extending from the foot-rope to both hand-ropes. This was among the negro race, so far distant from the Aryan people of the high plateau of Central Asia that it is difficult to understand how, even in the long course of history, there could have been any contact with or communication between them.

---

[4] Described in *Merry's Museum,* a serial magazine for young people, more than sixty years ago. Recollection verified by a contemporary reader. Encyclopedia Britannica, Fourteenth Edition, Vol. 18, Article "Primates," pictures the "spider monkey of tropical America, with long arms and long prehensile tails." These would be capable of the performance described.

[5] Appleton's Cyclopedia, Vol. XIII, Article, "Palm."

[6] Webster's International Dictionary.

[7] *Travel,* February, 1931, p. 38.

[8] *National Geographic Magazine,* Vol. 59, February, 1931, p. 231.

In Tibet, rivers are crossed by a single rope which serves, by aid of a trolley block and sling, and a dragline, to transport singly a man, an animal, or a parcel across a stream. (Type (2).)

The toboggan type of bridge (Type (3)) was also described by Mr. Kingdon-Ward.[7] This is a single rope securely anchored to a post or tree higher up than the point of landing on the opposite side of the stream. On this, slides a trolley block from which is hung a sling to support the weight of the traveler. The rope and the block are lubricated by yak butter—a product of Asiatic cattle. Standing in the sling, with his hands grasping only the block, the traveler is started by a push, and the force of gravity rushes him across the torrent with fearsome rapidity. Of course, another rope is set for crossing in the opposite direction.

The roadway type (Type (4)), shown in Fig. 1, needs little description. This example is in the Island of Java—Dutch Colonial territory. Apparently,

FIG. 1.—HIGHWAY TYPE OF ROPE BRIDGE, JAVA.

the span is about 100 ft., and the construction is quite like that which an engineer of to-day would make it with the material at hand.

The tubular type (Type (5)), shown in Fig. 2, was another found by Mr. Kingdon-Ward among the fierce Abor people on the Assam frontier. It spans the Dihang River, and is nearly 800 ft. long, the center being 50 ft. above the stream. Report was that 200 men built it in about a month, the labor being freely supplied by the villagers who used it.

A variation of the roadway type of bridge is described by Spanish historians as found by the invaders of the Sixteenth Century in the Peruvian Andes.[9] These bridges were composed of four ropes of twisted vines, covered with branches for a floor, and two ropes for hand-rails. To cross comparatively still water several hundred feet wide, the Inca workmen laid two great ropes of twisted grass on the water surface, on which cross-fascines of dry rushes were lashed. Two other ropes were laid on the fascines, which were again covered with smaller fascines, making a floating bridge with a floor, 15 ft. wide, over which armies passed. Such a structure required renewal or repairs every few months. The Incas also used a single rope on which an

[9] Edinburgh Cyclopedia, 1832, Vol. 4, p. 471.

under-slung hammock was held by clews so as to be drawn back and forth by a dragline. The widespread use of such "prehistoric" bridges in both hemispheres is significant of antiquity.

A French historian maintains that types of bridges found among backward tribes or communities in the hinterland of more advanced civilization

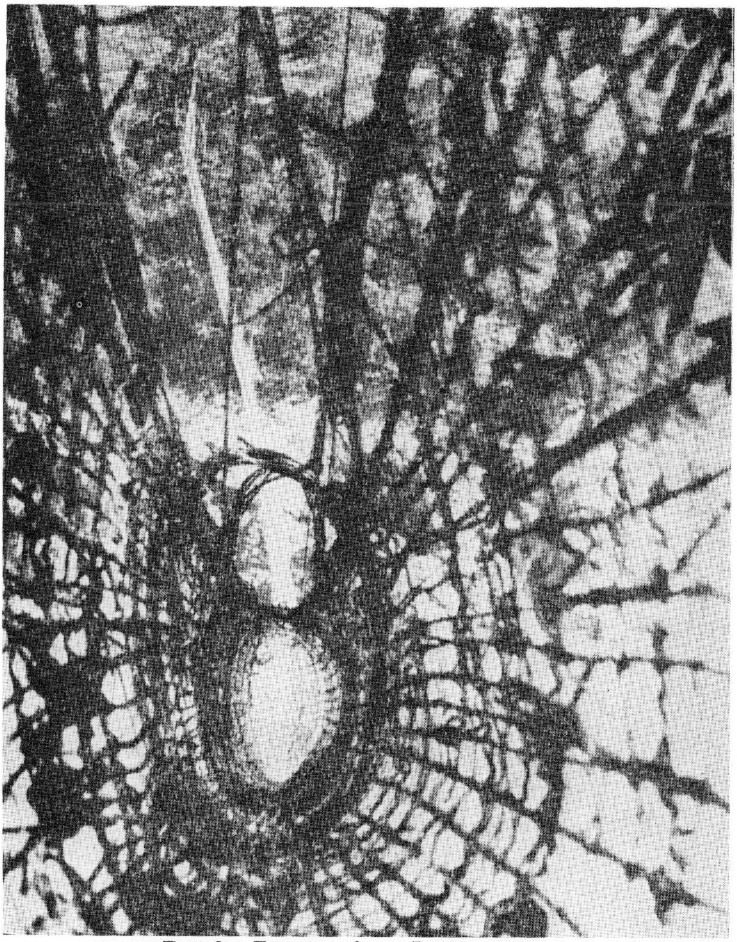

FIG. 2.—TUBULAR CANE BRIDGE, ASSAM.

are survivals of prehistoric structures with little if any change.[10] In the hills of Savoy in Italy, a primitive bridge was found, the structure of which had changed very little since the time of Julius Cæsar, who described[11] similar structures used by the Gauls, thus:

"It is a timber bridge or empilage, piled together rudely, not constructed by art. It needs no carpentry. * * * On each bank of the stream a rough

[10] "A Book of Bridges," by Brangwen and Sparrow, Plymouth, England, 1914 (in Dartmouth College Library).

[11] So stated by the author quoted.

foundation of water-worn boulders was laid, about fifteen feet square; upon this a criss-cross (crib-work) of the tree trunks was built so that the logs in the direction of travel, in alternate layers, were made to jut out farther and farther over the water, narrowing the gap to be bridged later by a few logs serving as beams."

Then, a floor of smaller round cuttings laid cross-wise over all made, in modern phrase, a corduroy road. Examples of such bridges in Kashmir, Northern India, and in Bhutan Province, north of Bengal, India, have been published by Charles S. Whitney, M. Am. Soc. C. E. One of these is apparently a development of what may be termed the corbel type into a real cantilever type of bridge. The portal towers, which probably serve partly as anchorages, suggest similar portals of Mediæval Europe. The total span appears to be between 80 and 100 ft.[12]

If prehistoric types survive to-day they would be among people like those of the "unchanging East," little affected by time or political disturbance, where the repairs made necessary by the vicissitudes of climate, decay, floods, and destruction by fire or enemies could be quickly and continuously effected.

The early historical period has little to tell about wooden bridges for many centuries. A writer states[13] that the Greeks, although masters of architecture, built few bridges and points out that "even Athens, with its splendid edifices, had no bridge over the Cephissus which ran athwart the most frequented thoroughfare to the city."

As to the Romans, Messrs. Brangwen and Sparrow state[10] that it has not been possible to learn how the Romans built their timber bridges. Pons Sublicius was a sacred monument, hallowed by traditions; but even this bridge was imperfectly described. According to the aforementioned writers, the most that can be said is that the Pons Sublicius was a tree bridge resting on piles. It dated from the time of Ancus Marcius who reigned from 640–616 B. C. Whether the chief priests did or did not build it, they were faithful in keeping it repaired, since it was in existence as late as the reign of Constantine (306 to 337 A. D.)—at least 900 years.

However, a reliable description is available of the timber bridge built (probably) under direction of the distinguished architect and engineer, Vitruvius, about 2 000 years ago, which permitted the passage of the Roman Army into Germany. This is frequently described as Cæsar's bridge. In his day, Palladio, the Venetian architect (1518-1580), made an exhaustive study of the vestiges of the Roman empire.[14] In his treatise on "Architecture," he gives, in the Latin, Cæsar's description of this bridge and a drawing to show the method of construction. The simplicity of the construction is obvious from Fig. 3. Its ingenuity appears by a brief study of the details, which shows how the parts of the bents lock themselves by action of the notched block in the notches on the cross-girder and inclined struts. Imposed weight simply makes the joints tighter. Thus, these bents could be framed

---

[12] "Whitney's Bridges," by Charles S. Whitney, M. Am. Soc. C. E., Pl. 31, p. 55; Pl. 32, p. 56, Edwin Wm. Rudge, Pub., 1929.

[13] Appleton's Cyclopedia, Vol. III, p. 270, Article, "Bridge."

[14] "The Architecture of Palladio," Large folio, with many full-page engravings; translated from the Italian, Lond., 1721.

quickly and set in place and could be removed quickly without the use of more elaborate fastenings, which would require time and supplies of hardware. It is on record that the Emperor made Vitruvius Superintendent of *Ballistæ* (Chief of Artillery), and this temporary structure in its design gives evidence of such skill as Vitruvius would doubtless possess. It is scarcely probable that the lumber was as smooth and square as the drawing indicates.[15]

A little more than a century later (104 A. D.) Trajan's bridge across the Danube River is mentioned in Roman history as one of the most prodigious works ever undertaken by the Romans. Dion Cassius stated that it rested on twenty piers, 150 ft. high and 170 ft. apart. These were united by semi-circular timber arches. Perhaps the height was fixed by the high banks of the river. Nicholas Bergier[16] states that the piers of squared-dimension stone

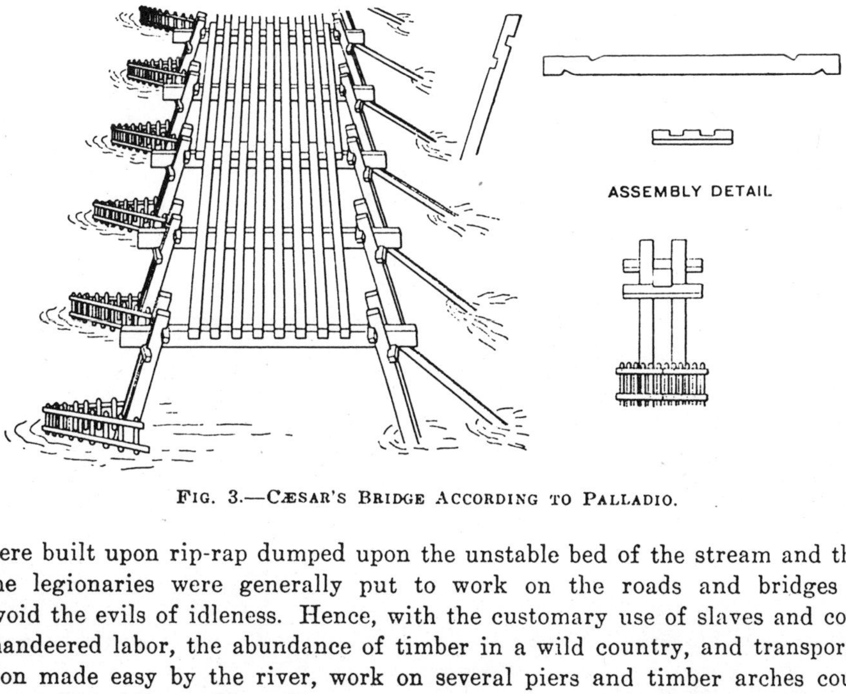

FIG. 3.—CÆSAR'S BRIDGE ACCORDING TO PALLADIO.

were built upon rip-rap dumped upon the unstable bed of the stream and that the legionaries were generally put to work on the roads and bridges to avoid the evils of idleness. Hence, with the customary use of slaves and commandeered labor, the abundance of timber in a wild country, and transportation made easy by the river, work on several piers and timber arches could proceed simultaneously. The carving, showing a span of this bridge, on Trajan's Column in Rome, as illustrated in books, is probably only conventional, since it is not in accord with the description given. Bergier presents long inventories of the roads of the Roman Empire in its greatest extent, from Britain to North Africa, and from Spain to Asia Minor, showing a total of nearly 65 000 miles. As to wooden bridges he states that timber was used in the early stages of road extension, but such bridges were eventually displaced by structures of stone.

[15] Appleton's Cyclopedia, Vol. XVI, p. 394; also, "Vitruvius," Encyclopedia Britannica, Fourteenth Edition, Vol. 23.

[16] "Histoire des Grandes Routes de l'Empire Romain," by Nicholas Bergier (2 vol.), Vol. 2, livre 4, Chapter 2, p. 610, pub. in Brussels, Belgium, in 1728. Nine chapters on Roman bridges.

Later European Timber Bridges

The Sixteenth Century.—The first illustration of a framed truss, in the modern sense, that the writers have yet discovered, appears in Palladio's "Architecture" (Fig. 4). That author states that this truss spanned a mountain torrent in Northern Italy. Apparently, the span is between 50 and 60 ft. The vertical members he terms pillars, or *collonelli*. The peculiar connections for carrying the weight from the cross-girder to the posts are possibly of his "invention," because he describes three "inventions," one of which is similar to Fig. 4, but with curved chords; a second, which has little merit, and a third, the framed arch shown by Fig. 5.

His designs seem to have attracted little attention because there was no further development of timber bridges in Europe until the middle of the

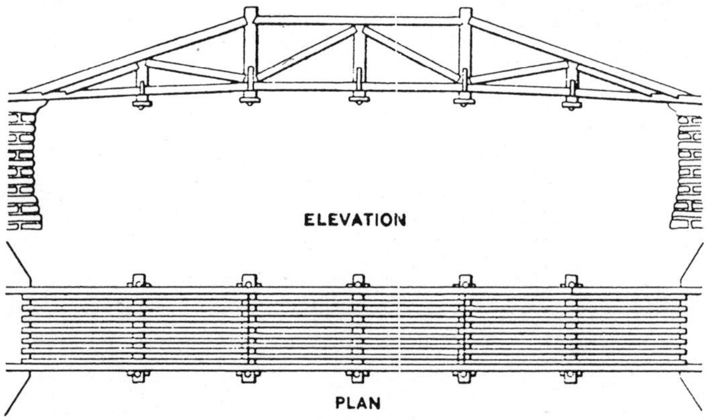

Fig. 4.—A Truss Dated About 1550 A. D.

Eighteenth Century—a period of rapid progress, when both rulers and people gave more attention to public works of all descriptions and when Civil Engineering began to be recognized as a profession. In this development French engineers took precedence, publishing elaborate treatises on archi-

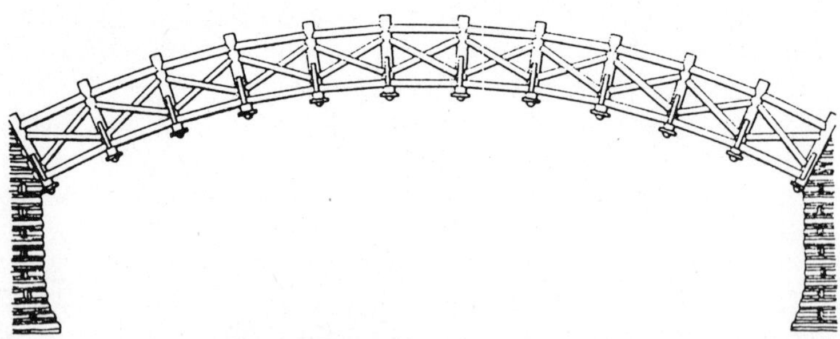
Fig. 5.—Palladio's Timber Arch.

tecture, carpentry, civil constructions, etc. In France, Perronet built his magnificent monumental stone bridges which served as models for English engineers and for those of other lands. He was a Director of "L'Ecole des Ponts et Chaussées," in Paris. English engineers, lacking abundant timber, utilized cast iron in some notable bridges standing to this day; but from the first they followed French practice in constructing some famous stone bridges.

The French types of timber bridges had the distinguishing characteristics shown in Fig. 6;[17] namely, level floors, low parapets, and flat arches, usually not framed after Palladio's pattern, but built of plank in layers clamped together. The floors were supported by radial struts attached to the arch and by braces from piers and abutments. The piers, usually of stone, were built at least to the skew-backs of the arches and sometimes to the floor level. The longer span shown (Fig. 6(*a*)) is about 150 ft.; the shorter one (Fig. 6(*b*)), is about 65 ft. In general, the designs produced pleasing archi-

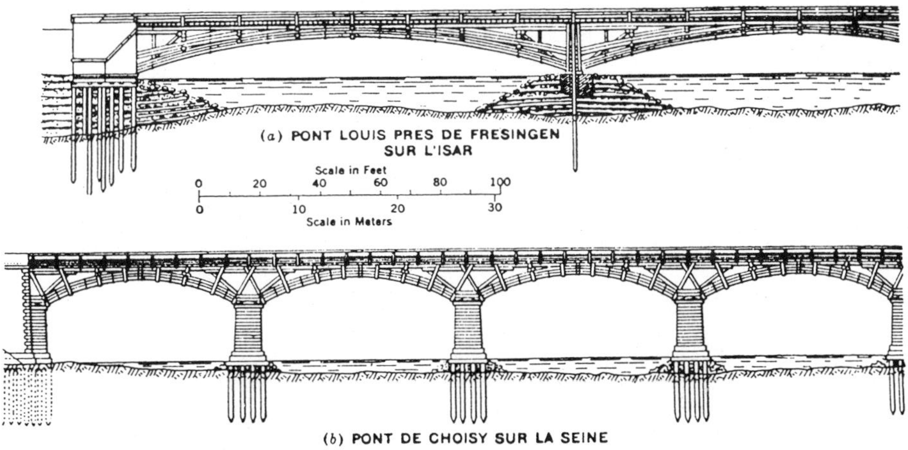

FIG. 6.—EUROPEAN TIMBER BRIDGES, EIGHTEENTH CENTURY.

tectural effects. Occasionally, piers were in the form of pile bents; and simple pile and beam trestles were resorted to when long spans were not needed.

Covered or roofed-in bridges were distinctly not a feature in European practice. Two notable Swiss bridges, built by the Grubenmann Brothers, were exceptional. The picture of the Schaffhausen Bridge has appeared in so many books and atlases as to be familiar to engineers generally. Fig. 7[18] shows the greater part of one span, with the details fairly distinct. It has been overpraised. It was a ponderous combination of long and short braces extending from as many points as possible to both abutments—a maze of timbers, scarfed, bolted, strapped, and clamped together to form nondescript trusses. The roof was top-heavy, with a needless amount of timber. It had two spans of 171 and 193 ft., propped by abutment braces. It was finished in 1758 and destroyed by the French in 1799. The one at Wittingen, Germany,[19] had a single span of 390 ft., making a stretched-out and awkward truss lacking the appearance of adequate strength. Neither bridge was worthy of consideration from an engineering standpoint. The assertion of an English

[17] From Gauthey's Atlas accompanying his 3-vol. treatise on "Constructions des Pont."

[18] "Notable Bridges"; Atlas of ten large folio engravings pub. between 1799 and 1825, by G. Taylor, 59 High Holborn, London, England.

[19] Gauthey's Atlas, previously mentioned.

writer that these bridges suggested the idea of the American covered bridge, with its roadway between trusses, is not justified by the facts. The early American types developed in the last two decades of the Eighteenth Century, and later, have little resemblance to those structures.

The only other European roofed bridges discovered[19] are: (1) The "Great Bridge" in Zurich, Switzerland, 128-ft. span, rather clumsily constructed with timbers laboriously scarfed from end to end and bolted together in pairs; (2) a bridge at Bassano, Italy, over the Brenta River, on pile bents in short

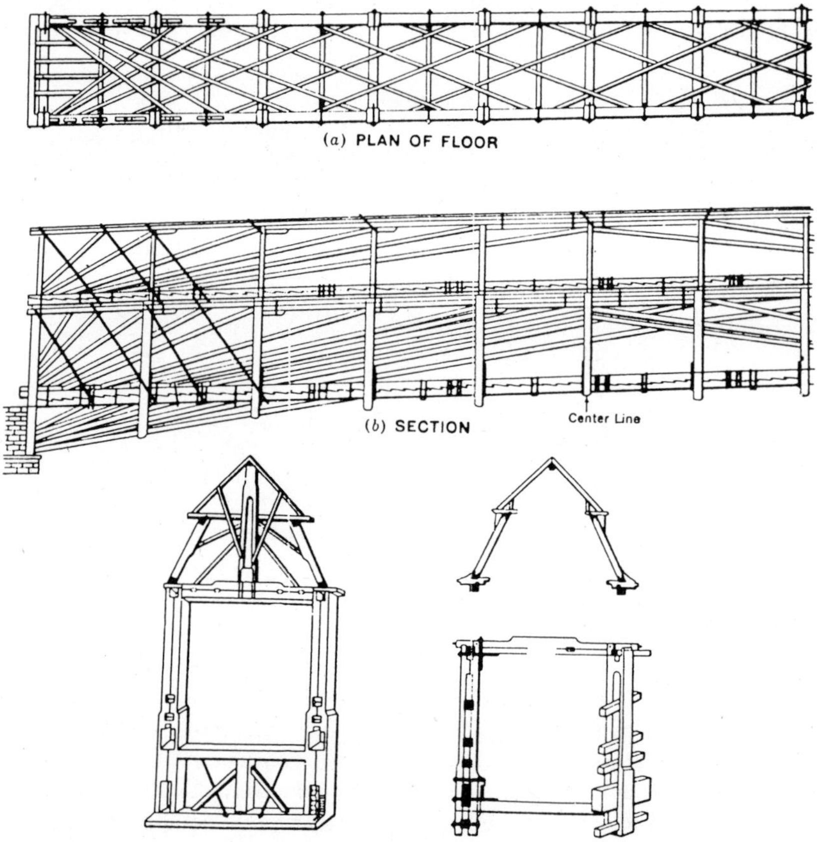

Fig. 7.—Plan and Sections, The Schaffhausen Bridge, Switzerland.

spans, and covered by an open colonnade supporting a light roof; (3) a project by M. Gauthey for a span of 50 m., which is impractical and not worthy of consideration; and (4) and (5) two other bridges, such as local carpenters might devise. The drawings do not indicate side covering.

American Timber Bridges

Development of Types.—The first settlers along the Eastern Coast of America found themselves in comparatively flat country, intersected by broad estuaries and tidal marshes. These conditions sometimes required bridges of considerable length and low elevation which were readily and inexpensively constructed by adopting pile bents and short spans. As loads were light, beams of moderate size were sufficient.

An authoritative history of this development was begun by the late Theodore Cooper, M. Am. Soc. C. E., in a paper presented to the Society more than forty years ago.[30] His short section on wooden bridges begins with the year 1660, thirty years after the settlement of Boston, Mass. He mentions "The Great Bridge" built in 1660, across the Charles River, which was 270 ft. long, on 13 piers. Twenty-five years later, another bridge was built over the same river, 1 503 ft. long, on 75 piers; the West Boston Bridge over the Charles River, 3 500 ft. long, on 180 pile bents, was finished in 1793; and in the same decade a pile bridge, a mile long, in 25-ft. spans, was erected over Cayuga Lake in New York State. In 1792, a bridge was built by Col. William P. Riddle over the Merrimac River, at Amoskeag Falls, now Manchester, N. H. This bridge was about 556 ft. long, supported by five cribwork piers filled with stone, and connected by beams not longer than 40 ft., which were then readily procurable from the great pine forests. Even before that, the Northwestern frontier had been pushed into Vermont, and a bridge was needed over the Connecticut River.

Colonel Enoch Hale's Bridge

In 1785, Enoch Hale built a notable bridge over the Connecticut River, at Bellows Falls, Vt. He was born in 1735 at Rowley, Mass., and moved to Rindge, N. H., in 1760, before it was incorporated as a town. He soon became prominent in the community, holding finally the office of High Sheriff of Cheshire County. In 1783, he obtained a grant from the Legislature to build a toll bridge over the Connecticut River between Walpole, N. H., and Rockingham, Vt. In 1784, he moved his family to Walpole and completed the bridge in 1785, two years after the close of the Revolutionary War. The bridge crossed the gorge at Bellows Falls, the narrowest point of the river where there is a rock reef dividing the water into two channels. This was the first bridge over the river at any point and while it was being built, the residents looked upon it as a foolhardy experiment. It is reported by contemporary newspapers to have been 365 ft. long, in two spans, supported at the center on the natural rock pier.

Fig. 8 is a sketch from an oil painting of the locality, showing the original structure, or a successor to it. The date of the painting is uncertain. Undoubtedly, the type shown represents that built by Enoch Hale. It consisted of four lines of braced stringers said to have had spans of 98 ft. each at the foot of the braces. The floor was at least 50 ft. above the water. It was the first bridge built in America with spans greater than could be negotiated with single stringer sticks; and, therefore, richly deserved the fame accorded to it.

[30] "American Railroad Bridges," *Transactions*, Am. Soc. C. E., Vol. XXI (July, 1889), p. 1.

The building of this first bridge over the river was noted far and wide. It was considered a remarkable feat of construction. It was reported in many newspapers[31] and was hailed as opening the Vermont region to the trade of Boston and the coast towns of New England generally. It was visited by

FIG. 8.—ENOCH HALE'S BRIDGE, BELLOWS FALLS, VT.

prominent men. The Rev. Timothy Dwight, President of Yale College, speaks admiringly of it in several of the narratives of his trips up the river. Without doubt the successful building and use of this bridge inspired the building of others in various sections during the next ten or fifteen years.

TIMOTHY PALMER AND HIS BRIDGES

One of the most ingenious and famous of the bridge builders of this period was Timothy Palmer, of Newburyport, Mass., greatly distinguished as a civil engineer (1751-1821). He may well be called the Nestor of American

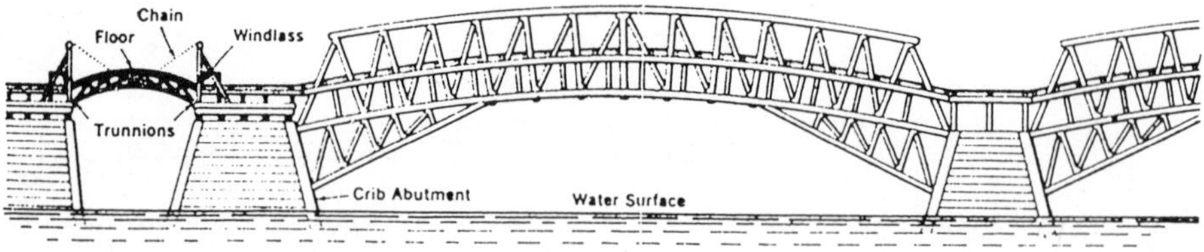

FIG. 9.—ARCH BRIDGE BUILT AT HAVERHILL, MASS., IN 1794.

bridge builders. The principal bridges on the Kennebec, Piscataqua, Merrimac, and Connecticut Rivers were built on his models. Mr. Cooper gives a view of the Essex-Merrimac Bridge, built by Mr. Palmer, three miles above

[31] *Massachusetts Spy*, February 10, 1785.

Newburyport in 1792.[22] It consisted of two arched spans, one on each side of Deer Island. The Piscataqua Bridge, built in 1794, had pile trestle approaches to an arch which was esteemed a wonder at the time, and the Haverhill Bridge, built the same year, has three arches so similar that Fig. 9[23] will serve to illustrate all three bridges. Mr. Cooper, quoting from a contemporary, describes[24] the Piscataqua Bridge ("The Great Arch") as a masterly piece of architecture. Its length was 244 ft., the rise was 27 ft. 4 in., and the depth of framework of the arch, 18 ft. 3 in. There were three concentric ribs, the middle one carrying the floor of the bridge. The ribs were made from crooked timbers, so that the fibers were nearly in the direction of the curves, and they were connected by pieces of hard and incompressible wood, with wedges driven between. The ribs were mortised to receive these connecting pieces and wedges, thus keeping an equal and parallel distance between them. Each rib was formed of two pieces, about 15 ft. long, laid side by side in such a manner as to break joints. Their ends all abutted with square joint against each other, and were neither scarfed nor mortised, the two pieces of timber being held together by transverse keys and joints. All the timbers were admirably jointed and freely exposed to the action of the air. Any piece might be removed for replacement without injury to the remainder of the structure. Another account states that, of the three ribs or tiers of girders forming the framed arch, the second or middle one supported the roadway, which was 33 ft. wide, and the third served for a railing.

An inspection of Fig. 9 shows a general conformity to the description of "The Great Arch." The short span of 30 ft. at the left is a draw-bridge over the channel. A feature of these bridges that should be noted, is the long braces at the ends, reaching from points half-way down on the abutments to points three-panel lengths forward on the truss.

Ten years later, in his "Permanent Bridge" (Fig. 10), built over the Schuylkill River (1804-1806), at Philadelphia, Pa., Mr. Palmer changed the

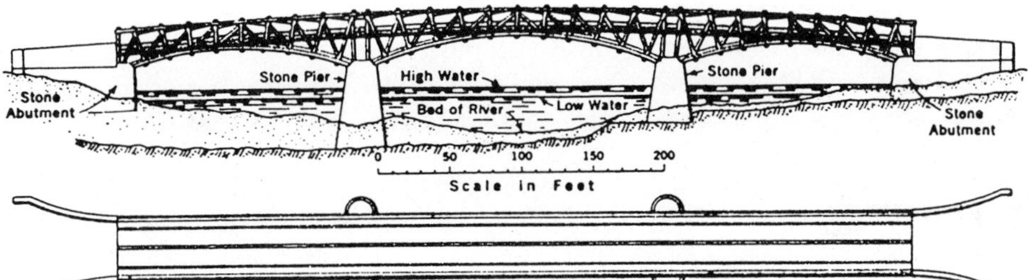

FIG. 10.—THE "PERMANENT" BRIDGE OVER THE SCHUYLKILL RIVER, AT PHILADELPHIA, PA.

type radically. The three spans, two of 150 ft. and one of 195 ft., are continuous over the piers, and there is no bracing beneath the arches. The floor is supported by a heavy continuous string-piece notched into the uprights of

[22] *Transactions*, Am. Soc. C. E., Vol. XXI (July, 1889), p. 4, Pl. I.

[23] This was copied from a pen-and-ink sketch made in 1797, now bound with some manuscripts in Boston (Mass.), Public Library.

[24] *Transactions*, Am. Soc. C. E., Vol. XXI (July, 1889), p. 3.

the continuous arched truss. The large timber available allowed deep notching and the tension members required projecting ends sufficient to resist shearing. Treenails were used instead of bolts and the timber was the best selected white pine. Meantime, in 1796, Mr. Palmer built a bridge over the Potomac River, at Georgetown, D. C., and secured a patent "for his improvement and construction of bridges." He insisted upon the necessity of protecting bridges from the weather, modestly stating that such precaution might extend the life of the bridge from ten to twelve years to as many as forty years. Whether his were the first of the covered bridges is not known.

His bridge over the Delaware River, at Easton, Pa. (Fig. 11), requires special notice as indicating Mr. Palmer's under-estimate of the durability of his work. The obvious changes from the type of the Schuylkill Bridge are the level floor, the straight upper chord, and the flatter arch. The details to be noted are the slotted bolts, the keys, and wedges. Treenails (commonly

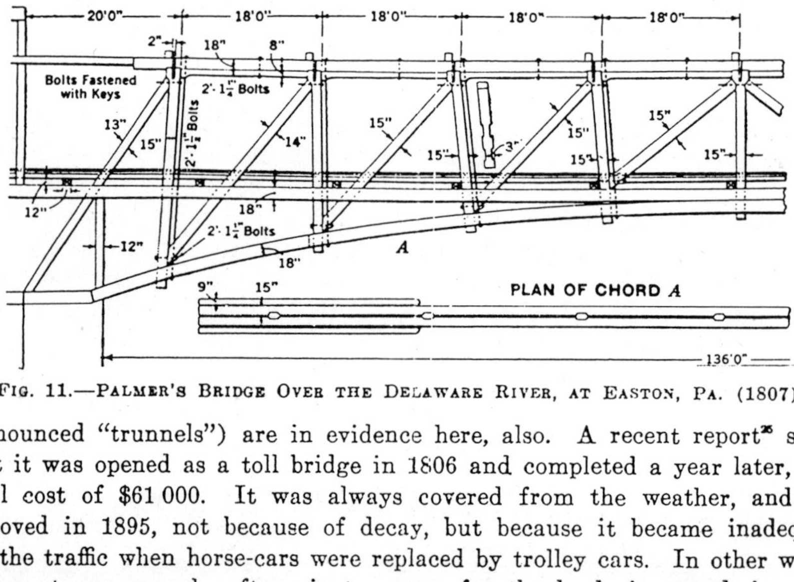

FIG. 11.—PALMER'S BRIDGE OVER THE DELAWARE RIVER, AT EASTON, PA. (1807).

pronounced "trunnels") are in evidence here, also. A recent report[25] states that it was opened as a toll bridge in 1806 and completed a year later, at a total cost of $61 000. It was always covered from the weather, and was removed in 1895, not because of decay, but because it became inadequate for the traffic when horse-cars were replaced by trolley cars. In other words, it was strong enough, after ninety years, for the loads it was designed to sustain.

In Fig. 11, attention should be directed to the style of connections. The bolts at the top of the posts are slotted and held in place by wedge-shaped keys or "fids." These bolts can be driven like drift-bolts in holes of the same size as the bolts, and can be "set up" or drawn in, as the timber shrinks, by re-driving the "fids," or inserting a washer if necessary. Hence, they serve both as treenails and clamps, while a drift-bolt alone serves only as a treenail, and a screw-bolt only as a clamp.

The Second Connecticut River Bridge.—Little more than a decade after Colonel Hale's bridge, a second bridge was built across the Connecticut River which was to be a link on the projected highway connecting Boston with

[25] Letter from State Highway Dept., Trenton, N. J., March, 1931. Information from Department records, through Roy Mullins, Highway Engr.

Northern Vermont. It joined the Towns of Hanover, N. H., and Norwich, Vt. Dartmouth College was influential in promoting the project. Professor J. K. Lord is authority for the information that[26] in 1796 one, Rufus Graves, built a bridge consisting of a single arch of 236-ft. span (similar to Fig. 9) at this point. The over-all length of the bridge was 344 ft., its width, 36 ft., and its stone abutments, 40 ft. square. It was arched to such a degree that the roadway at the center of the bridge was about 20 ft. higher than at the ends, presenting thus a sharp ascent on entering and a corresponding downward pitch on leaving. It was built of the largest selected pines, 60 ft. long, many of them hewed 18 in. square. The bridge fell by its own weight in 1804. On the testimony given by President Dwight, of Yale College, and others, it seems to be quite probable that Graves' bridge was patterned after the Piscataqua arch, but the builder lacked Palmer's experience and skill.

Lewis Wernwag and His Bridges

Lewis Wernwag was born in Germany in 1770. He died at Harpers Ferry, Va., in 1843. A letter written by his son, John,[27] to the late Samuel L. Smedley, M. Am. Soc. C. E., includes a list of twenty-nine bridges that Wernwag built during the twenty-seven years of his active career in Pennsylvania, Maryland, Virginia, Kentucky, Ohio, and Delaware. He was an original genius. His most noteworthy accomplishment was his "Colossus," built in 1812 over the Schuylkill River at the then, "Upper Ferry," later, Fairmount, Pa. This bridge, with a clear span of 340 ft. was composed of five parallel arched trusses, each with a rise of 20 ft. The roadway, however, had less than one-half that rise. Apparently, he originated the type. This bold design, scientific and architecturally beautiful, probably was never surpassed in America. One Swiss bridge has been mentioned, with a longer span by 50 ft., but compared to this, it was a monstrosity without merit. The cross-section shows the two roadways and two footwalks.

Wernwag did not take out a patent until 1829 and his drawing[27] shows iron rods in every panel, which were intended to act partly as counters and

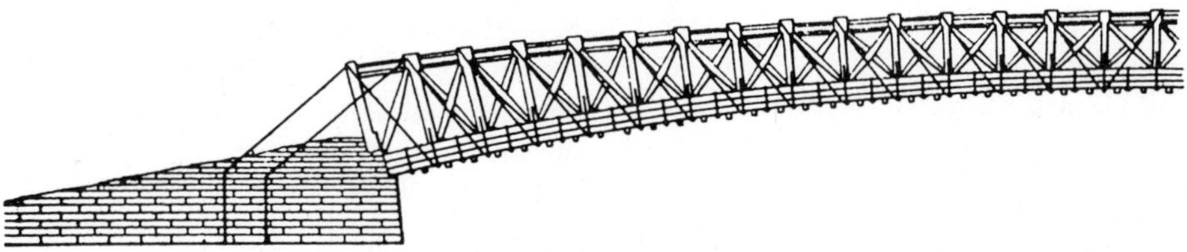

Fig. 12.—Longitudinal Section, The Upper Schuylkill River Bridge, Philadelphia, Pa.; "The Colossus."

also to adjust the joints if they worked loose. The comparatively light traffic loads of those days would cause but little reaction against the dead weight of such a heavy bridge. The arch was built up in seven layers, and had a total depth of 3 ft. 6 in., and a width of 13 in. Fig. 12 is from the patent drawing and Fig. 13[28] is a full span view of the bridge.

[26] "History of Dartmouth College," by J. K. Lord, p. 657.
[27] *Engineering News*, Vol. 14, (1885), p. 99.
[28] Copied from a large atlas of "Notable Bridges," pub. in London, England, in 1825.

In describing the bridge, Wernwag himself states that dry-rot was prevented by sawing the timbers through the heart to discover defects. The timbers were kept apart by iron links and screw-bolts and were neither mortised nor tenoned, except at the king-post and truss ties. No timbers

FIG. 13.—THE "COLOSSUS" OVER THE SCHUYLKILL RIVER, AT PHILADELPHIA, PA. (1812).

were allowed to come in contact, and provision was made to tighten necessary bolts as the timber tended to shrink, and any piece could be replaced without injuring the superstructure. The trusses were stayed at the ends by two long iron rods, one attached at the top of the end-post, the other at the top of the second post. Both extended downward at an angle of about 45° and were anchored at the back of a long and massive abutment built with squared-dimension stone laid in parallel courses.

In 1813-14, Wernwag built a bridge of quite a different type over the Delaware River, at New Hope, Pa. This had trusses with parallel chords, vertical timber posts, and iron rods for diagonals, anticipating what was known later as the Pratt type. The principal feature was the arches which rested on top of the abutments. In other respects, it was based on the same principle as "The Colossus" (Fig. 13) with the addition of the floor chord connected with the casting in which the arch rested, which prevented it from shoving against the pier or abutment.

In 1810 Wernwag constructed a bridge over Frankfort Creek between New York, N. Y., and Philadelphia, which was named "Economy." It was a cantilever[20] type designed so that the center panel could be tipped up in order to pass masted vessels through.

THE BURR TRUSS

Many famous bridges of various designs were built by Theodore Burr, of Massachusetts, during the first two decades of the Nineteenth Century. Fig. 14 illustrates a truss bridge built for railroad service over the White

[20] *Engineering News*, Vol. 14, 1885, p. 99.

River, at White River Junction, Vt. The photograph was taken when the bridge was being removed in 1890, to be replaced by an iron bridge. It was a splendid example of the Burr type, built in 1848 of old growth white pine when large sticks were available; the end-posts were 18 in. square. The hanger rods were inserted after the bridge was built, to assist in carrying

FIG. 14.—BURR BRIDGE OVER THE WHITE RIVER, WHITE RIVER JUNCTION, VT. (1848).

the floor load to the arches as engine loads became heavier. The masonry of this bridge was all quarry-faced granite ashlar laid with very close joints without mortar. At the right end is shown a corner of the north abutment. The abutment wings are circular arcs tangent at the face of the abutment, to a line parallel to the track, presenting the resistance of an arch to the pressure of the earth-fill. The bridge was as strong and serviceable after fifty-four years as when it was built. The entire combination—substructure and superstructure—was capable of much longer service than it rendered. Mr. Burr's first bridge was erected in 1804 over the Hudson River between Waterford and Lansingburgh, N. Y. (Fig. 15). This was followed by a bridge over the Delaware River, at Trenton, N. J., completed in 1806, which carried two roadways and two sidewalks by five arches;[30] the Mohawk River Bridge (1808), at Schenectady, N. Y., which Mr. Cooper

[30] *Transactions*, Am. Soc. C. E., Vol. XXI (July, 1889), Pl. VII.

described[31] as a wooden suspension bridge; and the long bridge over the Susquehanna River, at Harrisburg, Pa., a wooden arch bridge of twelve spans completed in 1816, on a radically different plan from that of the Trenton Bridge.

These four bridge types seem to have been experimental, in a sense, with Mr. Burr. His patent, dated 1817, which included a drawing of the Waterford Bridge type shown in Fig. 15, ultimately claimed nothing but the arch combined with the truss, the arches springing from the face of the abutment below the bridge seat. Although this bridge was built a year or two earlier, it looks as if it might be a derivation and improvement of Palmer's Easton Bridge (Fig. 11). The parallel chords of the trusses made ideal supports for the roof and floor, its posts and other members were firmly connected with the arches at each panel point, thereby acting perfectly as stiffeners for

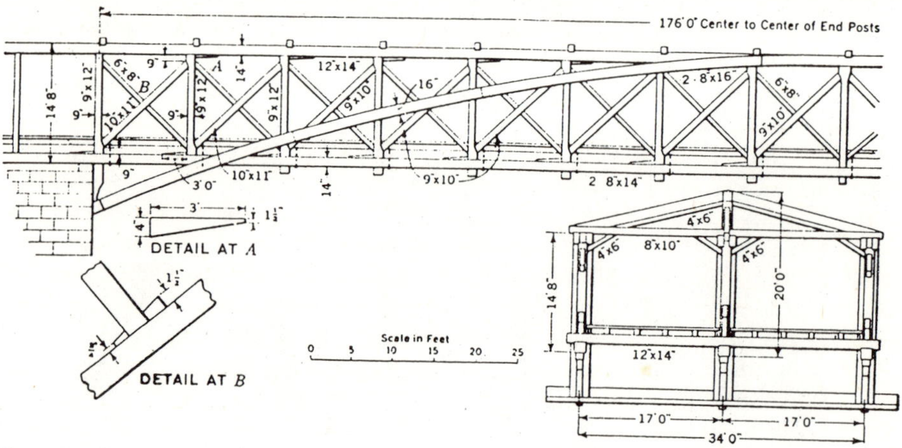

FIG. 15.—HIGHWAY BRIDGE OVER THE HUDSON RIVER BETWEEN WATERFORD AND LANSINGBURGH, N. Y., BUILT BY THEODORE BURR.

the arches. The bridge presents a combination of features the theoretical value of which was little realized by Mr. Burr. The other types, built with splendid workmanship, did commendable service, except the Mohawk River anomaly which had to be held up by a trestle after twenty years of use. The Waterford type, under the name of the Burr truss, was used all over the United States as long as purely wooden trusses were built. The type is well adapted for deck bridges, but few prominent instances of such use have come to the notice of the writers. In this type, Mr. Burr really inaugurated the combined arch and truss which has developed in Twentieth Century steel bridge practice, through the spandril-braced arch into the great braced arches at Hell Gate and Kill van Kull, in New York Harbor, and at Sydney, New South Wales, Australia. What are they but two-ribbed arches with stiffening trusses built between?

The Waterford Bridge, previously mentioned, consisted of four spans of from 154 to 180 ft. clear. It did duty for 105 years without material

[31] *Transactions*, Am. Soc. C. E., Vol. XXI (July, 1889), Pl. VIII.

strengthening until 1909, when it was destroyed by fire. The large timber available and the relatively light loads on bridges in this period of dirt roads rendered these bridges appropriate for the time, although they were frequently designed quite unscientifically. The posts were hewn, not only to make seats for the main diagonals, and for bearings on the arches, but were notched at the lower chord intersections. They acted in tension and had to be framed accordingly. The arches were 4 to 8 in. thick and were attached to each side of each truss. In case of unusually long spans concentric arches were used.[31a] Hundreds of highway bridges framed more or less (generally less), on the Burr principle, were built in various parts of the Eastern and Middle States and in New England. Many of these are still standing. Their longevity is remarkable, if properly protected by roof and side covering. The siding should leave a vertical space at the top about equal to the overhang of the roof; it should be vertical and nailed at the bottom to a girth furred out from the chord to leave a space so that dirt will not collect on the chord. Spans longer than 50 ft. should be reinforced by arches.

The timber in these early bridges was generally white pine which was a factor in their longevity. One at Springfield, Mass., was 104 years old when removed. Two over the Ammonoosuc River, one at Woodsville and the other at Bath, N. H., built in 1828, are still standing (1932). One at Greenfield, Mass., built in 1840, carrying all the traffic from the south, bound for the Mohawk Trail, survived in good condition until June, 1932, giving way then to a new bridge on a revised location. Many of the picturesque old covered bridges still dotting the New England landscape are of the Burr type, most of the shorter ones being without arches.

The longest single span of framed wooden bridge (262 ft.), since Wernwag's "Colossus," of which the writers have knowledge, crossed the Delaware River at Narrowsburg, N. Y. (1859). A system of struts between the posts below the upper chord to resist the horizontal thrust of the main braces, was an unusual feature in a Burr truss.[32] Attention is called to the unusual application of four separate and parallel arches in each of the trusses.

THE TOWN LATTICE TRUSS

Ithiel Town, of New Haven, Conn. (1784-1844), was granted a patent on a plank-lattice design for bridge trusses in January, 1820. It represents a lattice web, more suitable for a fence than for a bridge truss. Town secured a patent in 1835 for double webs and secondary chords.

A double-web truss of this type, with secondary chords, is shown in the paper by Mr. Cooper,[33] but it lacks much in the line of bracing and end support for the short lattice. It shows an unusual and unnecessary prolongation of truss over the abutment. It is true that this type requires a wider seat on the masonry than the Howe and Pratt types, in which the entire reaction is concentrated at a single panel point; but the additional masonry required for the lattice type is negligible if the parapet and wings are adjusted properly. Mr. Cooper, and other writers, state that these lattice-truss bridges are

---

[31a] See Fig. 14.
[32] *Engineering News*, Vol. 42, 1899, p. 122.
[33] *Transactions*, Am. Soc. C. E., Vol. XXI (July, 1889), Pl. X.

unsatisfactory on account of warping and getting out of line. This cannot be wondered at if lower lateral and effective transverse bracing are omitted, as is the case with the drawings usually shown. The scant justice done this type of truss by writers heretofore, is the warrant for recording its merits and according it considerable space in this paper.

Railroad freight traffic in the United States has become so enormously heavy that its economic handling requires equipment so ponderous that wooden truss bridges have become obsolete, even on branch lines, because rural industries must transport their supplies and products in the same equipment that is used on the main lines. On account of the common neglect of rural highway bridges it is generally advisable to recommend concrete structures where facilities for assembling materials therefor are reasonable; but where suitable lumber for lattice bridges is abundant and where fabricated steel is costly, as was the case in New England during the Nineteenth Century, and is now the case in Southern Alaska, Northern Russia, and Siberia, and perhaps in some parts of South America, lattice-truss bridges are economical and can be made perfectly good up to spans of 200 ft. for railroad trains not exceeding Cooper's E-40 loading. If properly covered and maintained, such bridges will give from 50 to 100 years' service.

As stated by Mr. Cooper, this truss has been built in many parts of the United States for railroads where suitable timber was plentiful and cheap. It has also been extensively used for highways before and since the advent of railroads in the Valleys of the Connecticut and Merrimac Rivers, in New Hampshire and Vermont. This and modifications of the Burr truss were the favorite types in building the picturesque old covered highway bridges, so many of which dotted the New England landscape and many of which still remain.

In 1900, at least 100 bridges of this type were extant on the Boston and Maine Railroad System.[34] They had been built largely on the "try-and-fail" method by local carpenters who adopted bridge building as their life-work. The best of these bridges, although built without any knowledge of unit strains, were quite satisfactory under the trains then using them. Common defects were lack of effective horizontal and transverse bracing, inadequate splicing of tension chord planks, and unsafe floors under derailed cars. Notwithstanding these defects, due to lack of scientific design, it is remarkable how well the "try-and-fail" method served, in the hands of the naturally competent mechanics of the Nineteenth Century, to produce bridges adapted to the loads and speeds using them. The excellent workmanship of Timothy Palmer's forces in building the "Permanent" bridge (Fig. 10) is typical of the work done by the early builders of these bridges.

Later practice on the Boston and Maine Railroad developed trusses designed more scientifically to overcome the lack of lateral and transverse bracing in the older bridges, as well as to provide for increasing weight of rolling stock. Fig. 16 shows details that have served to keep these structures straight and upright for thirty years and more, although subjected to loads far heavier than those for which they were designed.

[34] *Journal*, Assoc. of Eng. Societies, Vol. XV, 1895, p. 31.

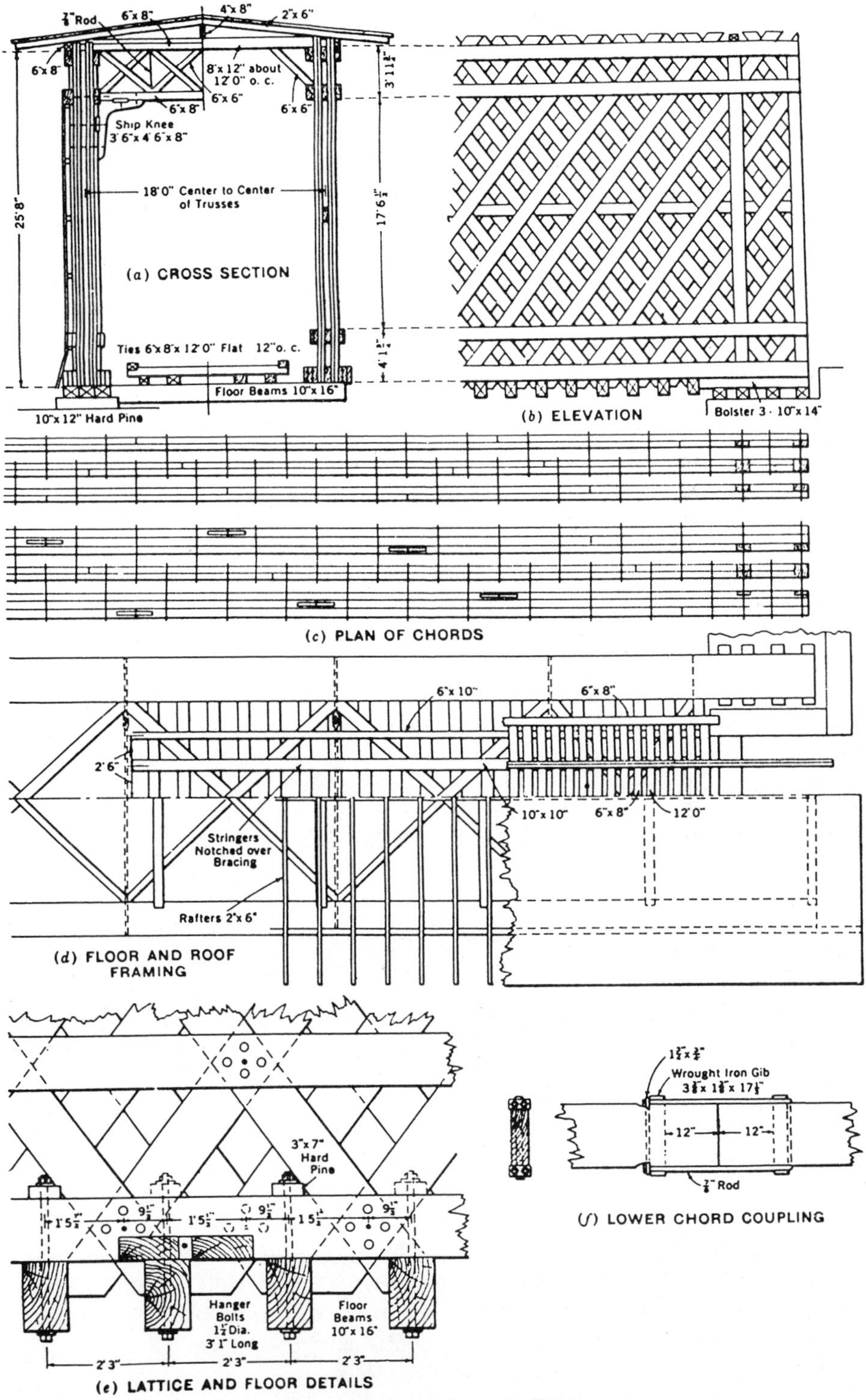

FIG. 16.—TOWN LATTICE TRUSS.

The distinctive characteristics of the Town lattice are its use of simple sizes of lumber, the small amount of framing required, and the need of nothing but bolts and a few round rods for metal work. In such regions as those mentioned, far from centers of skilled labor and steel fabrication shops, the simple character of materials and labor required by this type is important. The trusses are generally built of uniform sections throughout. While this is a feature that is often criticized as wasteful of material, such waste is much more than balanced by the resulting simplicity of framing and erection.

The Town lattice principle is similar to that of the English iron riveted lattice. Both will stand more abuse from service than any other type of truss. Both will give indications of distress long before collapse, and those that were properly built are found doing duty far longer than many other types. The most successful early builders of wooden bridges placed much emphasis upon the necessity of protection from the weather by roof and side covering. All later experience has proved that bridges not so protected deteriorate rapidly.

Until about 1890, lattice bridges could be built with spruce lumber (then costing about $18 per 1 000 ft. B. M.) for one-half the cost of iron bridges. Twenty years after that date (1910), steel bridges could be built as cheaply as those of wood; and now (1931) steel construction is much the cheaper. Computation for designs of this type is simple, following the methods commonly used for computing plate girders.

The accepted construction and a few important details are shown in Fig. 16. The timber ordinarily used in New England has been spruce, with the exception that for the longer spans Southern long-leaf pine was used for lower chords. The pins were of oak, 2 in. in diameter; and it was considered good practice to soak them in lubricating oil. Steel pipes are now being used in British Columbia for these connections. The stringers are gained over the lower lateral braces and rods. The upper laterals are 6 by 6-in. braces and $1\frac{1}{8}$-in. rods in nearly square Howe truss panels, with independent cross-beams and a transverse tenoned brace near each panel point. Turnbuckles in all lateral rods assist in keeping the bridge in order. The floor-beams are spaced to cut the ends of the tension diagonals less than those in compression. To secure sufficient strength in the lower chords of bridges without arches it has been found advisable in some cases to use iron couplings, as shown in Fig. 16. These are computed to add 16 900 lb. to the section if built as shown.

Noteworthy Bridge Builders.—Mr. Town was a promoter and salesman rather than a builder like his predecessors Palmer, Wernwag, Burr, etc. He sold rights to build his design, published advertising pamphlets, made a trip to Europe, and exhibited business ability of high order.

Sanford Granger (1796-1882), of Chesterfield, N. H., established his home at Bellows Falls, Vt., and became a contractor for bridge and building work. There is a tradition that he patented the lattice type of bridges, but it is more probable that he acquired territorial rights from Ithiel Town to build them. He organized a crew for building this type of bridge and constructed many in Southwestern New Hampshire and Eastern Vermont.

Nicholas Powers (1817-1897), of Clarendon, Vt., during the middle of the Nineteenth Century built many bridges over a wide section of country from Maine to Maryland. His was the famous Blenheim, over Schoharie Creek, in New York State, described subsequently. As Superintendent of Construction he built the great railroad bridge at Havre de Grace, Md., over the Susquehanna River; and as Designer and Builder, many of the covered bridges in Vermont and vicinity were due to him.

Dutton Woods (1809-1884), born in Henniker, N. H., adopted bridge building as a life work in 1837 when railroad construction began to be active. He erected many bridges of various types as desired for several railroad companies from New Haven, Conn., to Concord, N. H., and farther north, as well as many highway bridges for various towns and cities. His favorite type in later years was the Town lattice. He used plank arches where needed. He held the position of Superintendent of Bridges on the Concord Railroad until his death.

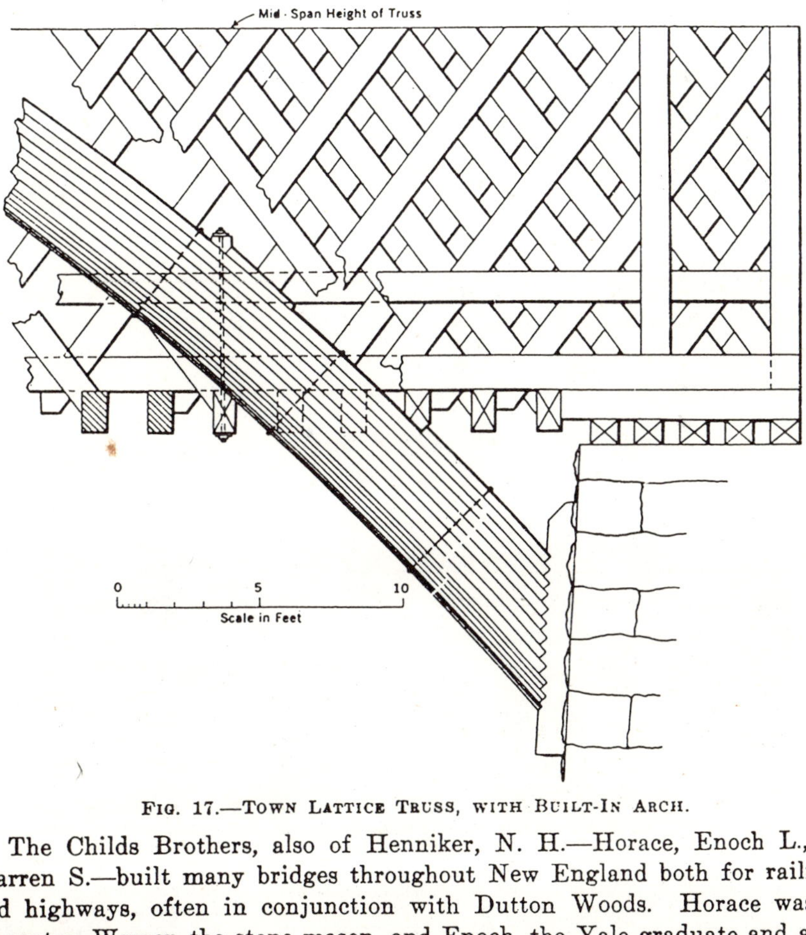

FIG. 17.—TOWN LATTICE TRUSS, WITH BUILT-IN ARCH.

The Childs Brothers, also of Henniker, N. H.—Horace, Enoch L., and Warren S.—built many bridges throughout New England both for railroads and highways, often in conjunction with Dutton Woods. Horace was the carpenter, Warren the stone mason, and Enoch, the Yale graduate and architect-engineer.

David Haselton, born in Dorchester, N. H., in 1832, deserves mention as a builder of lattice bridges after 1860. He was Superintendent of Bridges and Buildings on the Boston and Lowell Railroad until 1890, when it was leased to the Boston and Maine.

Supplemental Arches.—Spans of more than 80 ft., depending on the load, should be built as combined truss and arch. Ordinarily, a single arch on the center line of the truss is sufficient, but, if desired, an arch may be built on each side of the truss, Burr fashion. These arches spring from skew-back blocks scribed to fit the masonry below the bridge seat. The arch planks are firmly held together by radial bolts, and the load is applied by vertical suspension rods carrying needle-beams below the lower chords, or by passing through the floor-beams if they are underhung. These suspenders (or posts when over that part of the arch near the end which is below the lower chord) should be not more than 10 ft. apart along the arch to prevent it from buckling on account of its laminated structure. Fig. 17 shows this construction.

Combinations of arch and truss have been disapproved by schoolmen because of the impossibility of determining the theoretical distribution of stress between the truss and the arch; but if there is strength enough in both, and they are properly yoked together, the safe elasticity of the wood and the connections will take care of the distribution. The Town lattice is the best of the various types of wooden trusses to serve as an arch stiffener because its web members serve either as counters or main braces. The value of arches framed with trusses, if heeled against the masonry, has been amply demonstrated by other types of trusses, some of which have stood for a century. If these arches are sprung from the lower chord the purpose of using them is defeated.

Highway Bridges.—Hundreds of highway bridges have been built with lattice trusses, mostly with single lattice; that is, with one web system per truss. Fig. 18 shows such a bridge of two spans built in 1840 and removed

FIG. 18.—TUCKER BRIDGE, BELLOWS FALLS, VT. (1840).

in 1930 to be replaced by a two-span concrete arch structure. This bridge replaced the successor of the famous Enoch Hale Bridge (Fig. 8), and was on the identical site, about 15 ft. higher than the old one. It failed from neglect of a decayed wall-block which led to irreparable injury to one truss. It was 262 ft. long over all. Tolls were eliminated in 1904.

The Timber Lattice.—The type of truss shown by Fig. 19 has been used extensively for highways in the Connecticut River Valley and especially in Vermont, and, in a few instances, for railroad bridges. It is not known who first adopted it,[34a] but its merits have been abundantly proved. The arches shown below the chord were added in 1927 to relieve a decayed lower chord.

FIG. 19.—THE LEDYARD BRIDGE, HANOVER, N. H. (1859).

Several highway bridges of this type built 60 or 70 years ago are now standing, some with spans of 160 to 180 ft., over the Connecticut River. The most common reason for unsatisfactory conditions in this type is decay of some part not effectively protected or not renewed in time. In common with plank lattice of their time, built without knowledge of unit strains, they are frequently weak in the tension chord and deplorably deficient in floor-beam strength. (Blenheim Bridge, a different type, shown in Fig. 20, is mentioned subsequently.)

Unlike railroad through bridges, the trusses of highway bridges are not of great height so that they seldom suffer from lack of transverse or lateral bracing. Their plank floors, unlike the open floors of railroad bridges, make the best kind of lateral bracing. The usual construction is a web composed of intersecting joists, 6 by 8 in., or larger, set with broad sides parallel to the plane of the truss, in panels of about 4 ft. which, with trusses of approximately 16 ft., out to out of chords and a run of three panels, make seven crossings of each web member in its length.

At these intersections each member is notched 1¼ in., making the web 9½ in. thick. Where the members pass through the chords, both web members and chord-sticks are notched 1 in. each, leaving 7½ in. of net timber for the web and 5½ in. clear between the sections of the chord. For spans of less

[34a] Ithiel Town obtained a patent on it in 1839, p. 332.

than 75 ft., secondary chords are not needed. The ends of the diagonals should be cut not less than 9 in. vertically from the chord, in order to allow a safe length to resist the longitudinal shear.

Pony Trusses.—For spans ranging from 30 to 50 ft., pony trusses have been used extensively. Modified Howe trusses with inclined end posts, Town lattice, and mongrel types are common. The trusses of this class must be stayed by sway-braces. It is not good practice to set these braces on an extended floor-beam because each passage of a load would throw the top of the truss back and forth on account of the deflection of the floor-beam. The needle-beams carrying these braces should be blocked down so that the stringers shall not rest on them.

FIG. 20.—BLENHEIM BRIDGE, SCHOHARIE COUNTY, NEW YORK (1855).

Stringer Bridges.—A history of wooden bridges would not be complete without mention of the humble stringer bridge which was, of course, the earliest recourse of man for crossing deep streams, as stated under "The Ancient Background." In American Colonial days just such structures with log cribs or rough stone abutments and round logs hewn on top to receive a plank floor were used for openings up to 40 ft. For wider streams pile or frame trestle bents were used in quiet waters and log cribs filled with stone where heavy freshets occurred. In the days of large timber and of dirt roads this type sufficed for highways, but with railroads, spans of plain stringer bridges were limited to about 15 ft. Fig. 21 shows a type of single-track, pile-stringer bridge of the last decade of the Nineteenth Century.

For spans from 15 ft. to 30 ft., compound keyed stringers have been found useful. Fig. 22 shows the details. The keys shown are cast iron and are proportioned for the longitudinal shear and, hence, the total depth of the compound stick can be used in computing its moment of resistance. They are placed as nearly as may be in accordance with the intensity of the shear. The thin fin at the middle of the key is cut into the wood to hold the key from working out of place.

Fig. 22 shows the junction of a compound and single stringer on a trestle bent. The pile-cap is called a "girder cap with rider"; it is used only on first-class pile work. For lower grade pile work and for framed bents, solid caps, either tenoned or drift-bolted to the pile or post, are used.

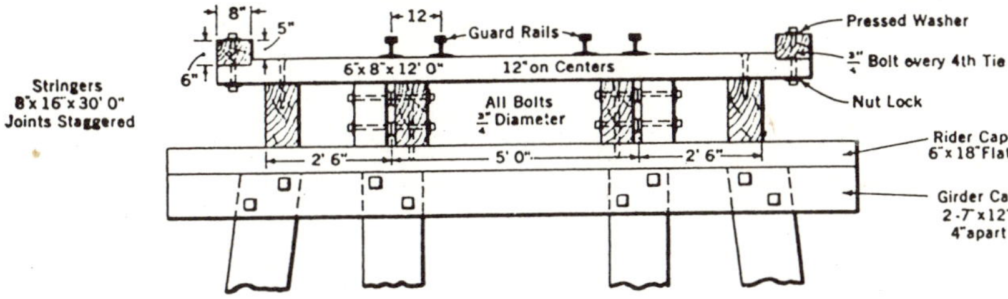

Fig. 21.—Standard Trestle Section.

Although it can be said that the day of wooden bridges has passed, it is a fact that many hundreds of miles of wooden trestles, stringer, and truss bridges are still in service in the United States. This brings the discussion down to the simple stringer and trestle, illustrations of which, other than Fig. 21, are scarcely needed.

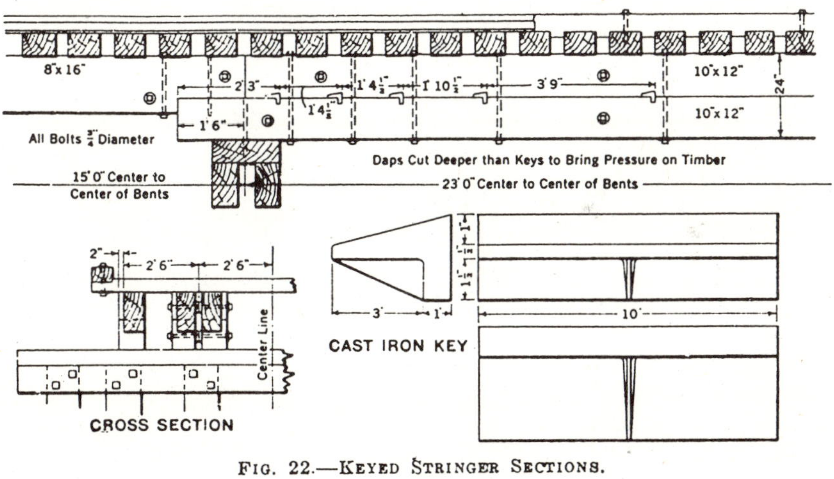

Fig. 22.—Keyed Stringer Sections.

Long's Bridges

In 1830, Bvt. Lt.-Col. Stephen H. Long, U. S. Topographical Engineers, patented a truss which bore his name. He afterward modified it and was granted other patents in 1836 and 1839, "to produce lateral stiffness or inflexibility in wooden or frame bridges." This was effected by a system of lattice bracing applied either to the upper or lower chords, or to both. Concerning Colonel Long's truss Mr. Cooper remarks[35] that this is another step in simplicity and concentration of parts. Here also iron did not enter as a necessary part. Connections were made by framing the parts together or by use of wooden keys or treenails. Many bridges of this form were built, but

[35] *Transactions*, Am. Soc. C. E., Vol. XXI (July, 1889), p. 9.

it was never widely popular. Fig. 23, from a drawing of 1830, shows two sets of braces at the abutments below the lower chord reaching forward to the first and second panel points.

The remark of Mr. Cooper applies in particular to the design of 1830. Colonel Long's later designs departed from the simplicity displayed by the first one. The use of abutment braces extending to the first and second

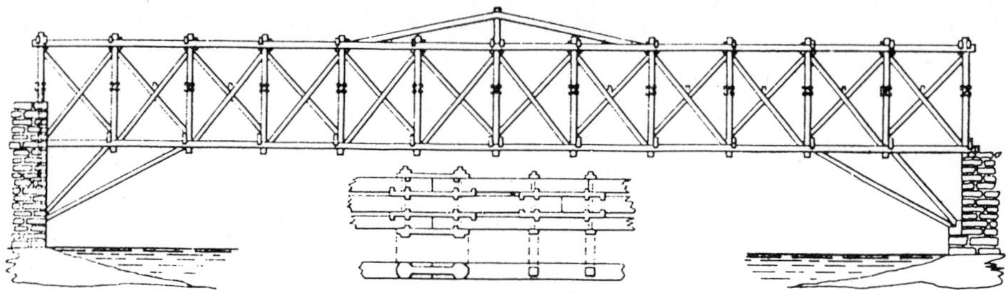

FIG. 23.—LONG'S BRIDGE.

panel points breaks the continuity of the frame as a simple truss, independent or unassisted between the bridge-seats. The term, "assisted truss," may be applied to it. The cogged cast-iron splice-blocks (or plates), which his patent drawing shows, might be used instead of wooden splice-blocks. This idea might have found favor in the days of cast iron, so to speak, but is more worthy of consideration now when steel castings with their greater strength are more readily available.

Many rural highway bridges were built throughout the Eastern States by local carpenters, which were hybrids of the Long and Burr types, without the science of Long nor the arches of Burr. It is impossible to classify them as of any type but, due to the practical common sense of their builders and the accommodating property of wooden framing to distribute strains, these "mongrel" bridges have often given service equal to some of the widely known patented types.

THE HOWE TRUSS

In 1840, a patent was granted to William Howe, of Massachusetts, which was the basis of the Howe truss bridge so generally built throughout the country for half a century and more after the date of the patent. There were two systems of web members; that is, the diagonals ran two panels. The patent drawing also shows a secondary lower chord about 2 ft. or more above the main chord, the only function of which seems to be to carry the horizontal thrust of the end brace of one system of diagonals. It is doubtful whether there was ever a truss built with this arrangement of secondary chord and short end braces. During this season (1840), Mr. Howe, in company with Amasa Stone (1818-1883), built the great bridge over the Connecticut River, at Springfield, Mass., to carry the Western Railroad (later the Boston and Albany), which was opened for traffic the following year. This bridge had seven spans of about 190 ft., center to center of piers. The trusses had two systems of web members like the patent drawing, but the end diagonals of both systems were carried on a single-angle block at the bridge seat. (See Fig. 24.) The lower chord consisted of six sticks, running six panels of about 7 ft. each, packed close together without splices of any kind. The diagonals and rods were of uniform size throughout. In 1874, the bridge

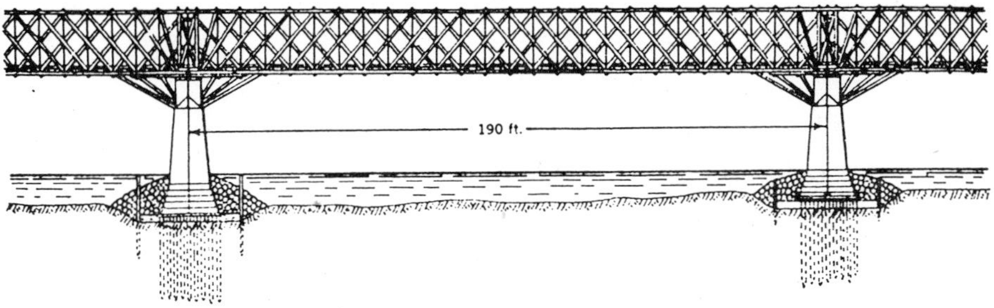

Fig. 24.—Bridge Over the Connecticut River, at Springfield, Mass.

was replaced with an iron riveted lattice. It is not known to the writers that there has been another bridge built of the Howe type with webs of two systems. In 1841, Mr. Stone bought the Howe patent, formed a company, and built this type of bridge in many parts of the country.

In this type the vertical iron rods are a vital part of the web system; and the possibility of proportioning the web members for the stresses which they transmit, as well as the simplicity of the joint connections through which that transmission is effected, make this a nearly perfect type of truss. Early bridges of this type were built of white pine and, later, of spruce, and, finally, of Southern long leaf pine which, although much heavier than the other species named, is better adapted for the Howe type. The splice-clamps and keys are usually of seasoned oak. Oak angle-blocks were used for thirty years or more, but later practice considered cast iron better than wood to transmit the forces acting at the panel point. In general, Fig. 25 shows, by isometric projection, the arrangement of the members of a Howe truss, deck, railroad bridge built in 1874 over the gorge near Quechee, Vt. By cutting off the members assembled at one panel point the view discloses all the truss members and auxiliaries in the bridge, namely, the upper and lower chords, the web system of the main and the counter braces and the vertical rods, the horizontal lateral braces and rods, the vertical sway-braces, the latticed lateral system nearly the same as that proposed by Colonel Long, wooden angle-blocks, and the common arrangement of splice-clamps.

This bridge had a pair of light arches (which is unusual with this type), taking load from the truss by means of vertical rods which, passing through the needle-beams below, received whatever load the latter transferred to it. A proper splice is shown in the lower right-hand chord, and an imperfect, one-sided splice is shown on the farther chord at *D*. Each chord-stick was spliced in every fourth panel. This bridge had a clear span of about 200 ft. and a height of 24 ft., center to center of chords. There were eighteen panels, $11\frac{1}{2}$ ft., center to center of verticals. The track was about 165 ft. above Ottaquechee Brook, and the bridge was carefully protected by boarding.

More recent practice improved the application of the Howe design by increasing the size of the web members and decreasing the strength of the lower chord-stick connections from the center of the truss to the ends, according to the stresses sustained. Fig. 26 shows a Howe truss substantially as built in 1889 at Morrisville, Vt., which displays these advances over the older practice. The rods in the panels next to the center are two $1\frac{3}{4}$-in. and one $1\frac{1}{4}$-in. and, in the end panel, three $2\frac{3}{8}$-in. and two $2\frac{1}{4}$-in., all upset, with gib-plates varying from 6 by $1\frac{1}{4}$ by 36 in., next to the center, to 14 by $1\frac{7}{8}$ by

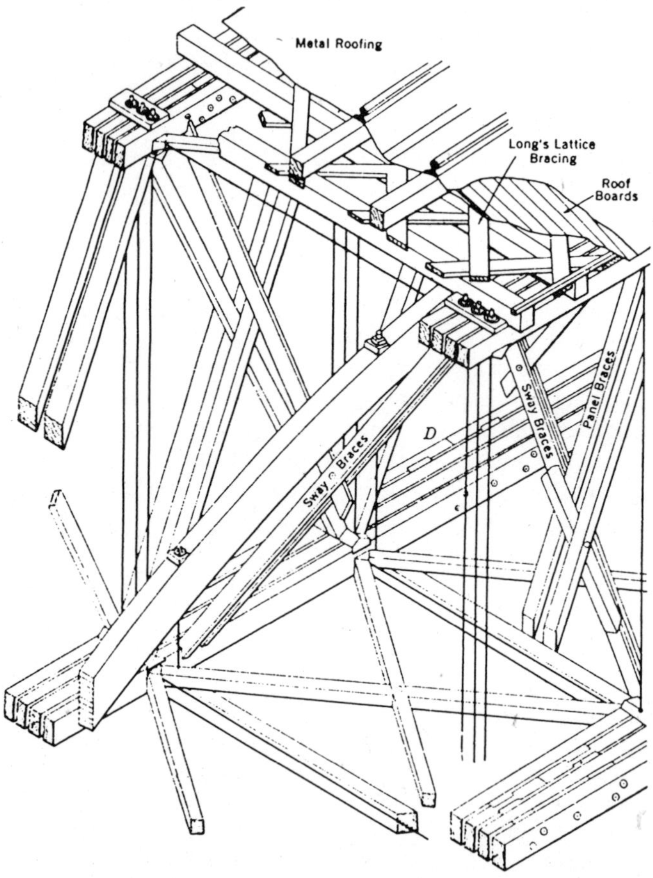

FIG. 25.—ISOMETRIC VIEW OF HOWE TRUSS ACROSS OTTAQUECHEE BROOK NEAR QUECHEE, VT.

50 in., at the ends. The main braces vary from three sticks, 7 by 7 in., at the center, to three sticks, 11 by 13 in., at the ends. The increase in strength of the lower chord-stick connections from end to center can be seen on the plan. Note the filler opposite the single splice-clamp in the fourth panel from the end. The cast-iron angle blocks are all of the size computed for the end panel. This practice leads to excess of iron, but saves expensive pattern-making.

For through bridges, in order to have sufficiently rigid portal-bracing and to return the side boarding to cover the end panel inside, the upper chords are extended to a connection with the end-posts which may well be in two sets with light rods between them, as shown. Howe truss bridges should be protected from the weather by roofs and side covering, the same as other types of wooden bridges. The lack of this protection has caused the premature failure of many otherwise good bridges, the life of which would have been prolonged two or three times if they had been covered as advised by the early masters of timber framing.

In 1850, a bridge was built for the Sullivan County Railroad Company, at Bellows Falls, Vt., across the Connecticut River, by George A. Parker, of Massachusetts, who, later, was Chief Engineer of the great Havre-de-Grace Bridge over the Susquehanna River in 1862. In the Bellows Falls Bridge were two 250-ft. spans of Howe trusses built with an arch on each side of each truss, Burr fashion. This bridge was replaced in 1882 by a riveted iron bridge on the same masonry that carried the wooden bridge. Arches, although sometimes built with Howe and Pratt trusses, are not indigenous with them so to speak. The south span was saved by its arches during a great freshet in 1862, when a span of the highway bridge at Charlestown was swept off and came floating down the river erect and whole, striking the railroad bridge "head on." An eye-witness states that, as the lattice-web system of the highway bridge crushed and gave way against the rugged chord of the railroad bridge, the ridge-pole of the "attacking" bridge pierced through the standing bridge, and protruded so far on the down-stream side that, as soon as the wreckage had quieted enough, men ventured out on the bridge and sawed off the overhanging stick for fear it would twist the truss. The bridge was crowded a few inches out of place on the abutment. No further damage was done.

A somewhat similar case, showing the usefulness of arches, occurred at Penacook, N. H., where a lattice bridge with an arch built on the center line of each truss (see Fig. 17) spanned the Contoocook River near its mouth. A great jam of ice, during a winter thaw, piled up against the bridge, several feet above the floor level, and sprung it 2 ft. out of line at the center; but the arches, seated on the face of the abutments, withstood the pressure, and no serious damage occurred. Except for the arches the bridge would have been pushed into the Merrimac River.

During part of the Civil War (until 1866), Mr. Parker was engaged in building the bridge for the Philadelphia, Wilmington, and Baltimore Railroad Company over the Susquehanna River between Perryville and Havre-de-Grace, Md. The superstructure was of the same type and span length as the Bellow Falls Bridge. The writers know of no other Howe truss bridges built with arches in pairs, as in these two.[36] Nicholas Powers, who had worked on the Bellows Falls Bridge, was asked by Mr. Parker to work on that at Havre-de-Grace, under a Mr. Bagley, the Superintendent. When the spans were nearly erected a tornado wrecked them, twisting an approach

[36] A good wood engraving of two of the twelve spans of the Havre-de-Grace Bridge resting on stone piers and abutments, may be seen in Appleton's Cyclopedia, Vol. III, p. 273, pub. in 1873.

trestle badly out of line. Mr. Bagley was blamed and discharged. Mr. Powers prepared to go also, but the Manager asked him to take charge of straightening the approach, and finally arranged with him to superintendent the reconstruction. Mr. Powers agreed to do it for $7 per day if, to start with, he could pick fifty men. His terms were agreed to, with the promise of $500 bonus if he finished the bridge without accident. Mr. Powers had his 16-year old son as Assistant, and, at times, had 400 men employed on the work. Meanwhile, the haying season was on at the home farm in Vermont. To the urgent letters from his wife to come home and secure the hay crop, he replied by sending money to hire the needed help. The bridge was finished sometime in November, a celebration was planned for the dedication, and a picture was to be taken of the bridge crew with Mr. Powers in the center; but he had promised his wife that he and their son would be home for Thanksgiving; he kept the promise and the picture was taken without him.

The great popularity of the Howe truss was well deserved. Many hundreds were built in all sections of the United States and Canada. However, this type cannot be classed strictly with wooden bridges, inasmuch as it uses iron rods for web-tension members—one of the four essential elements of the open framed truss. The development of railroads at the time of Mr. Howe's patent (1840), gave a great impetus to the manufacture of iron, and made it possible to procure round rods in quantity. The type is really intermediate between wooden and iron bridges.

Iron angle-blocks were sometimes cast with tubes extending through the chord, as in Piper's patent of 1867, for the double purpose of furnishing resistance to horizontal thrust by being notched into the chord-sticks; and, also, by contact with the gib-plates, to resist the vertical pull of the rods. Some writers specified that the tubes be made short enough to allow for shrinkage of the chord timbers; but such shrinkage may be more or less than the allowance; and it is best to provide sufficient hold on the chord itself for the horizontal thrust, and also gib-plates large enough to prevent side-grain crushing of the chords. If the tubes and gib-plates come in contact, the chord is generally loose and does not properly support the angle-block. Hence, it is best to omit the tubes and design the angle-blocks and gib-plates to do their part without them. Unless the cast-iron angle blocks are unusually large, it is not profitable to attempt to save metal by coring the central part, although this is the practice of some designers.

Colonel Long's patent of 1830 claims a cast-iron splice plate for the chord-sticks. A modification of this as a steel casting for splicing the chord-sticks in Howe trusses is well worthy of attention. The plate can carry four cogs instead of two, as shown by the wooden splice, and can act as a key by having cogs on both sides and also as a spacer if rests are cast to bring the plate in the center between the spacing of the chord-sticks. The principal difficulty in the use of these plates is in boring the holes if the timber and plates are in place; but it can be done by the skillful use of a ship-auger type of bit (see Fig. 23).

Designing the angle-block of the Howe truss involved the brace-in-panel problem, which is to have the squared ends of the timber brace and the

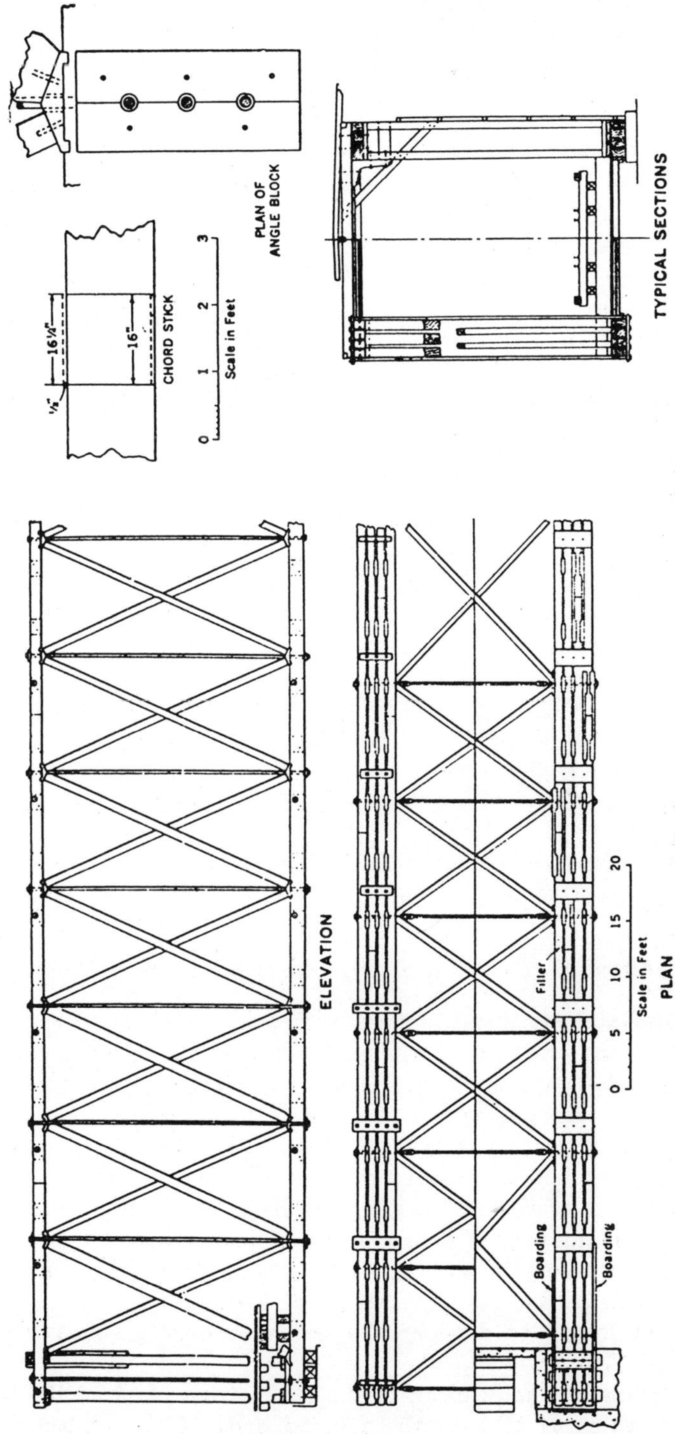

FIG. 26.—PLAN AND SECTIONS, HOWE TRUSS RAILROAD BRIDGE (1889).

inclined bearing surface of the angle-block fitted so that the center line of the brace and the vertical axis of the rod will meet on the center line of the chord. This involved also the camber to be given to the truss. In theory, it is a problem in higher mathematics. In practice, each builder usually had his own rule, and the best one was to lay out a panel full size and cut a template for the brace, leaving just such an angle-block as he wanted, and then, for camber, cut his counters a little short, the amount determined by his experience. The camber was given partly for the sake of appearance. Optically, to some observers, a truly horizontal chord might appear to have a slight sag. Even an actual slight sag was of no consequence if it did not increase so as to indicate over-strain of the truss. However, there is a good reason for camber, because of the compressibility of wood and the imperfection of hand-made joints.

Keys and joint clamps should be framed with bearing surfaces and notches cut with a slight draft so as to require driving to place when erected (see details in Fig. 26) and the depth of the notches should be such that the end fiber resistance to the compressive strength at the notch should be equal theoretically to the tensile resistance of the uncut cross-section of the necks of the clamps. The notch must be made far enough back from the end of the chord-stick so that the lengthwise shearing resistance of the softer wood shall be so adjusted to the pull that a slab will not split off between the notch and the end. This is usually the weakest part of the connection. Many splices have failed by such splitting. Almost invariably, in old bridges of this type, outside sticks have only their inner side clamped, as shown at *D* in Fig. 25; and all the clamp bolts are in the head of the clamp, leaving the ends of chord-sticks free to bend outward, which increases the tendency to split off at the notch. The builders were very loath to bore through the neck of the clamps.

The Pratt Truss

A patent was granted in April, 1844, to the late Thomas W. Pratt, M. Am. Soc. C. E., and Caleb Pratt, the claims for which are indicated in a drawing of which Fig. 27 is a copy. In this patent the attempt was made to

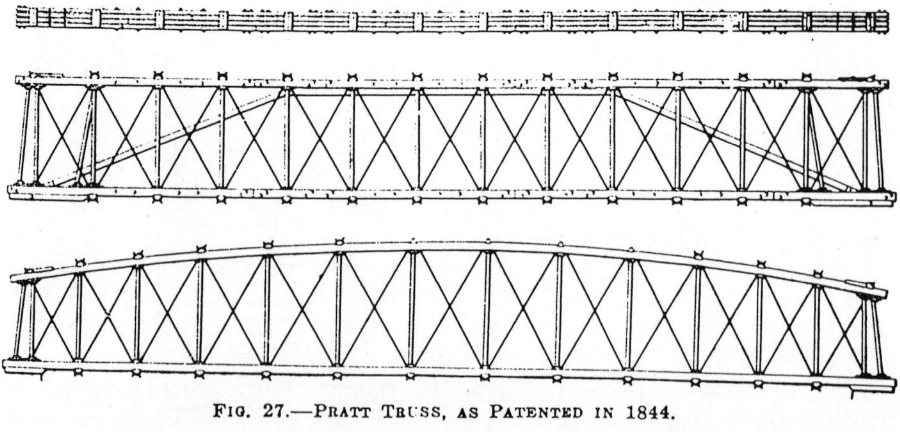

Fig. 27.—Pratt Truss, as Patented in 1844.

use timber in the simplest and most efficient manner in the web members by confining it to the verticals. The advantage of doing this, however, is over-matched by the longer rods required than in the Howe truss, and the awkward angle-blocks needed for these diagonals. For all metal trusses, however, the Pratt truss is ideal.

For railroads, the continuing increase in weight of equipment and loads has kept the problem of strengthening bridges before engineers. For the early Howe and Pratt types it has been found that the tension rods are the weakest single feature. In the Howe truss, yoking on additional rods is an easy remedy; furthermore, reinforcement is easily applied to the timber diagonals. In the Pratt truss, this is far more difficult; but the Pratt truss permits curved upper chords, as shown in the lower drawing, and this is desirable for long spans. Evidently, in the Howe type, curvature of the upper chord would introduce complications in making proper seats for the varying lengths and inclinations of the diagonals.

In the parallel-chord truss the king brace and strain-beam shown in the upper drawing of Fig. 27 is a reversion to the European practice more than a century old in Pratt's time. In a panel truss it befuddles computation of web stresses and is a faulty device on several counts for assisting a truss which should be designed to be strong enough without it.

The activities of the Childs Brothers in building bridges, especially for the New England railroads when these were young, has been noted herein previously. The drawing accompanying the patent granted to Horace Childs in August, 1846, indicates an attempt to combine the Howe and Pratt types. It is defective in omitting counter-rods in the middle panels, and peculiar because of an awkward attachment of the main braces by a joint bolt passing through the chords and the angle-block, and up into the brace, which was a crude method of making the main braces act as counters. The Childs Brothers built many bridges of various types, among them the Granite Street Bridge, in Manchester, N. H., over the Merrimac River, in 1862, and one over the same river in Hooksett, N. H., in 1858.

Toll Bridges

Incidentally, the term, "toll bridge," has been applied to several of the old bridges previously described. Some may not realize that, at first, with possibly a rare exception, all bridges were built by private initiative under charters granted by the State. The charters specified the tolls to be charged, subject to change only by decree of the Court. In populous districts the interruption of travel at the toll-gates became so inconvenient and annoying that counties and towns bought the charter rights and gradually made the bridges "free."

Of thirty bridges built across the Connecticut River between New Hampshire and Vermont, north of Massachusetts, beginning in 1785, twenty-six were chartered by law and operated as toll-bridges. At present (1932), all but two are free.

In this Twentieth Century, comes a reversion to the toll system, or policy. In the great metropolitan districts, and at points where wide streams must be crossed by trunk-line highways, the cost of erecting and maintaining the huge and lengthy structures, with capacity sufficient for the enormous volumes of light and heavy traffic, is too great to be collected by general taxation. It must be retrieved by toll charges upon those who use the facilities.

Wooden Draw-Bridges

The roads over the broad estuaries and tidal streams mentioned under "American Timber Bridges," invariably required openings for the passage of water craft. In pioneer days, land transport was difficult, and it was desirable to get water-borne traffic as far inland as high tide would float the small vessels then used. Invariably, draw-bridges called "tip-ups" were built in such cases. They were narrow so that four oak logs sufficed for stringers. Crooked trees were selected with large butts which helped as counterweights. These logs were 12 to 15 ft. longer than the width of the channel to be spanned. The stringer portion over the opening was hewn as small as practicable. An oak log firmly framed to the stringers, with rounded ends to turn as trunnions, served the purpose of a hinge. Chain guys passing over sheaves at the top of a gallows frame spanning the roadway, actuated by hand windlasses, raised the free end of the bridge when desired. If the opening was too wide for a single leaf, two were used meeting at the center, with a toggle arrangement to hold them in grade under passing loads.

These bridges required the lifting of a section of the plank over the hinge, to allow the draw stringers to turn up. This is simple in a highway floor; but with the introduction of railroads, difficulty developed in lifting the rails and their supports, so that a swing bridge of the type called the "**A**-frame swing-draw," was adopted. The stringers of this bridge were framed to rest at the heel—when swinging—on a pivot under the center of the track, at a distance back from the channel cap. The head of the **A**-frame was vertically over the heel pivot when hauled backward for raising the swing end of the bridge. The guy rods running from the head of the frame to the needle-beam under the stringers were held apart for side clearance of the train by a spreader-beam high enough above the track for proper head clearance. A hauling-off beam attached by a pin to the side of the bridge carried a rack into which a pinion, operated by a hand-crank on shore, meshed to swing the bridge off and on,[87] the same as indicated by the illustrations referred to under "The Jackknife Draw-Bridge."

The Jackknife Draw-Bridge.—The **A**-frame draw was awkward to arrange for in some locations. When the Gloucester Branch of the Eastern Railroad was built in 1845, a draw-bridge of some other type was required for the entrance of the little harbor basin at Manchester, Mass. Joseph Ross (1822-1903), a young mechanic of Ipswich, Mass., and Samuel Ashburner, a draftsman, engaged on the work, "got their heads together" and designed what was afterward called the jackknife type of draw-bridge. This important development has been adequately described elsewhere.[88]

---

[87] *Engineering News-Record,* November 5, 1931, p. 719

[88] *Loc. cit.,* January 21, 1932, Vol. 108, p. 93.

A Unique Bridge

A bridge of singular distinction, the pride of Schoharie County, New York, deserves special mention. The location is North Blenheim, on Schoharie Creek. (See Fig. 20.) Erected and long maintained as a toll bridge, it was finally taken over by the town which has carefully kept it in prime condition. The record states that the "contract to design and erect the bridge was awarded to Nick Powers, a Vermont Yankee who built many such structures," and that "the bridge was not built over the stream as many would believe, but was constructed piece by piece, back of the present village near the site, and there set up to make certain that every part was perfected before erection over the river was attempted." While this was being done, the masons built the abutments. The material is native oak and pine. The unique feature is the single arch, framed and built on the center line of the middle truss, and this truss extends upward 6 or 7 ft. higher than the side trusses, so that its upper chord forms the ridge of the roof. If the floor timbers extended across the entire 26 ft. of the width, thus serving both roadways, they would contribute greatly to the stiffness and stability of the entire structure.

From available evidence it seems likely that the posts and diagonals were framed to pass through the chords as in Long's design. It has the unusual span of 228 ft., but this gives no distinction over bridges already considered. It can scarcely be classified or correlated with any other type of bridge that has come within the knowledge of the writers. Built in 1855, it is claimed to be quite serviceable for the traffic which it now carries after seventy-six years of usefulness; but the County Officials decided that it must be replaced by a modern concrete structure. However, according to later report, the bridge will be maintained as a local antique.

Retrieving Weak and Injured Bridges

It may be said that it is more feasible to strengthen a wooden bridge in a satisfactory manner than one of metal or masonry; and that the deficiency in the strength of a bridge scientifically designed is more difficult to remedy than in one built by guess, when one member is proportionately weaker than the bridge as a whole. Timber bridges are subject to decay and various accidents, and may develop structural weakness of parts, all of which require timely attention.

First-aid or emergency treatment involves, generally, the setting of trestle bents under the bridge. Where these cannot be permanent, many bridges may be strengthened by the addition of plank arches inside the truss, if side clearance permits. Vertical hanger rods that pass through needle-beams hung below the chords carry the bridge load to the arches.

Lattice bridges are sometimes struck by running ice during spring freshets, and the lattice plank projecting below the lower chord may be broken off, so as to weaken, seriously, the power of the lower chord to carry the floor-beam load. Hanging the lower to the second chord by simple rods or yokes is the usual remedy.

If bridges tend to get out of line, as some do, they are guyed with wire ropes or rods to well-fixed bolts or other permanent anchorage on the banks of the stream. Decayed or broken parts may sometimes be renewed in kind. A common fault in all types is insufficient strength in the floor system. This calls for additional floor-beams or stringers, or both.

Lower chord-stick connections in all types are often deplorably weak in panels near the middle of the span. Fig. 25 at *D* shows a defect exceedingly common. The pull of the single clamp tends to throw the end of the stick outward, which greatly aggravates the tendency of the head to split at the notch. The remedy for this is an additional clamp cogged in and bolted on the outside; or, if split, by various types of iron couplings shown in Fig. 28, or in Fig. 16, for use in a 4-in. plank. The couplings can be extended to

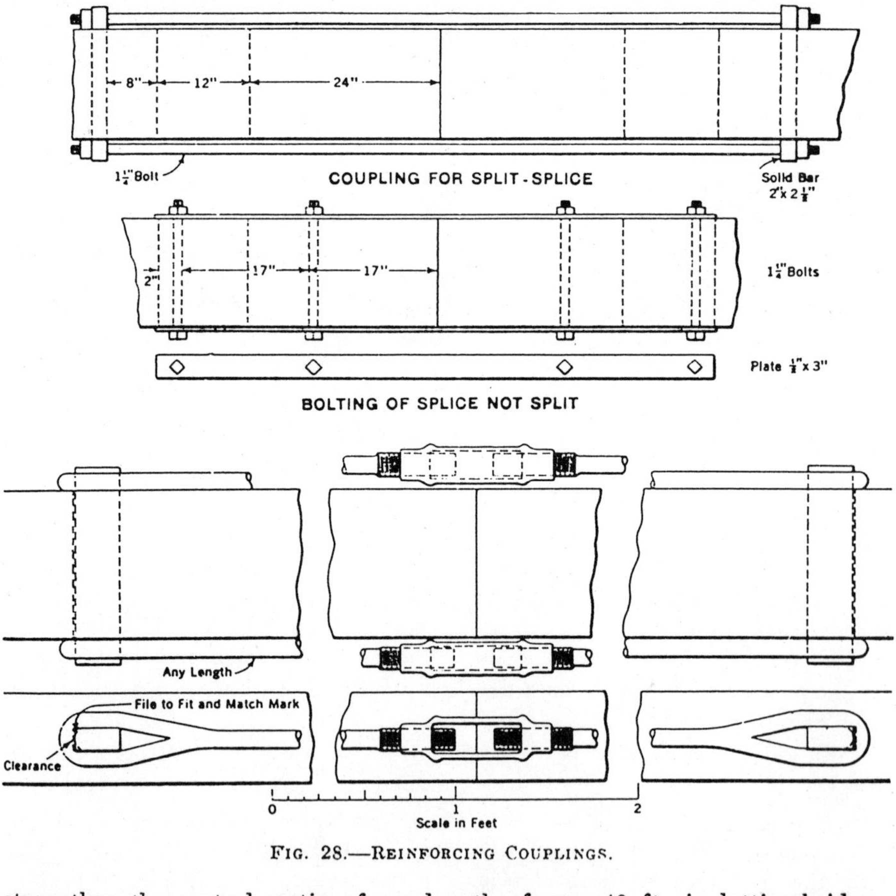

FIG. 28.—REINFORCING COUPLINGS.

strengthen the central section for a length of, say, 40 ft., in lattice bridges, using loop ends at the gibs and right and left-hand threads for a turnbuckle at the center. Multiple rods can be used in this manner without trouble. Another favorite method is to attach an auxiliary stick to the side of the chord, connected with effective keys and large bolts or pipe drift-pins.

The tension rods in Howe and Pratt trusses are frequently found to be deficient in strength. Reinforcement is made by yoking a pair of rods to the truss with gib-plates of wood or iron across the chords, as near the panel points as possible. Reinforcement is more difficult in the Pratt than in the Howe trusses on account of fitting the gib-blocks to resist the horizontal pull of the inclined rods. The web rods in most of the older bridges were not upset, but facilities for upsetting are now such that its neglect in the long rods required for through railroad bridges is inexcusable.

The bridge builder cannot be too careful as to designing the bolted and spliced joints, in order to equate properly the stress and resistance where the parts come together. Manifestly, the tensile strength of coupling bolts and bars, the beam strength of gib-bars and bolts, the compressive resistance of cut-fiber surface and side grain of timber, and the tensile strength of the uncut section of the pieces united or joined, are all factors to be considered. Adjustment should be made so as to secure, with reasonable accuracy, equality of action and reaction throughout the combination.

Bridge Curiosities and Atrocities.—An interesting sidelight of the subject of timber bridges is the example, now and then, of the effect of ignorance and carelessness on essential details of construction. Three letters on this subject were published[39] in 1893.

Summary

Reviewing in succession the advancing phases of this history, the writers have discussed the twisted-vine-cable bridge, in various adaptations to existing conditions, as the natural conception of primitive and early barbarian people. Then, in less barbarous and more semi-civilized times, the overhanging timber cantilever structure followed as an extension of the plain crib-work and log stringer.

Around the Mediterranean Sea, during the period of the colonizing and commercial activities of the Phœnicians and Greeks, there seemed to be little thought of bridges. Land commerce was borne by camels across the deserts, and by other pack animals in more settled regions. Wheeled vehicles were, for the most part, war chariots. The then abundant forests provided the timber for the ships sent forth from Tyre and Sidon. Ezekiel, the seer, wrote that the ship planks were made from fir trees, the masts from cedars of Lebanon, and the long oars from the oaks of Bashan.[40] A greater requisition on the forests was made later when the Greeks, Carthaginians, Persians, and Romans, in their long-prowed biremes and tiremes, fought the historic naval battles for the mastery of the inland sea.

It is conceded now that the conception of the arch as an architectural feature in masonry structures originated in the Near, or Farther, East. Quite recently archæologists have uncovered a tomb in Ur of the Chaldees, entered under a true arch of burnt brick. This is said to be 5 000 years old. In Europe, the extensive use of timber for bridges doubtless began when Rome

[39] *Engineering News*, July 20, 1893, p. 60; September 14, 1893, p. 219; and November 9, 1893, p. 376.

[40] Ezekiel, XXVII, 5 and 6. About the Seventh Century B. C.

first started to develop its expanding system of military and post roads. Cæsar's Rhine Bridge is a fine type of emergency bridge which would be creditable to a modern engineer. That the early Roman timber bridges embodied the preference for the semi-circular arch appears from the description of Trajan's Danube Bridge. After the long lapse of the Dark Ages, when the Renaissance brought new interest in public works, the flat arch became the standard type. It was expressed in open framing and bracing, such as proposed by Palladio; or solid timber arches, all usually with nearly level floors and side parapets, with no protection from the weather. The ponderous framing of some Swiss bridges like that at Schaffhausen, with their awkward lengthy bracing, found little acceptance. The low arch types, usually examples of good architecture as well as good engineering, prevailed during the Seventeenth and Eighteenth Centuries, until the later European timber bridges were displaced, during the Eighteenth and Nineteenth Centuries, by the splendid masonry bridges which stand as enduring monuments to distinguished engineers.

Since the aboriginal Americans had little use for products of the forest, the earlier American settlers along the coast had plenty of timber for pile and stringer bridges. Then Enoch Hale made courageous utilization of the magnificent white pine timber at his disposal for bridging the Connecticut River at the Bellows Falls gorge. Following him, the skillful and resourceful Timothy Palmer, and the ingenious and conscientious Lewis Wernwag built their great flat arches of wood somewhat stiffened by truss framework on top. Their insistence on careful protection from the weather inaugurated the distinctly American type of covered "through" bridge. In the first decade of the Nineteenth Centry, Theodore Burr brought out his idea of combining arches springing from the face of the masonry below the bridge seat, with a real truss, the lower chords of which carried the floor. This type of truss perhaps inadvertently on the part of Mr. Burr inaugurated a new era in bridge building, rendering it possible to trace the actions of loads from the point of their application to the supporting masonry.

At the end of the second decade Town brought out the lattice type; and, at the end of the third decade—in 1830—Colonel Long patented his design. These three types, Burr's, Town's, and Long's, are the only important ones, extensively used, that were practically constructed wholly of wood. From 1830, the building of railroads gave a great impetus to the building of bridges; and, in 1840, the Howe type with iron-rod vertical web members, became a favorite; and, in the next decade, the Pratt type, with iron-rod diagonals and timber verticals, soon led to iron lower chords; and then followed the so-called combination bridge interspersed with the early all-iron structures.

It is interesting to note the growth of the use of iron in bridge building. The building of clipper ships along the New England Coast up to the middle of the Nineteenth Century developed there the art of framing heavy timbers, the fastenings of which were made chiefly by treenails ("trunnels"). The iron used was fashioned by the smiths in the shipyards. Timothy Palmer, trained in those yards, used some slotted iron bolts, tightened by keys, as can be detected in Fig. 11. Use of threaded bolts became general afterward. Town used treenails only, but later practice in his type of bridge prescribes bolts freely, and rods where supplementary arches are used. In 1840, Howe made iron-rod verticals the basis of his claim, while Pratt, in 1850, extended their function as diagonals, and also made the curved upper chord a feature. The combination type called for iron diagonals and lower chords with compression members of timber. As early as 1840 all-iron bridges were built with cast-iron compression and wrought-iron tension members. The ingenious and scientific Squire Whipple, Hon. M. Am. Soc. C. E., enters the field here, using this combination of cast and wrought iron successfully for many years. In 1859, Howard Carroll built the first all wrought-iron bridge for railroad use. It was a riveted lattice truss, and the contest between the riveted and pin-connected types was on. In the last decade of the Nineteenth Century steel took the place of iron as the best material for bridges; and now, ninety years after Howe's introduction of iron as an integral part of the truss, other alloys of steel are coming to replace carbon steel in the art.

ACKNOWLEDGMENTS

Special acknowledgment is due to John W. Storrs, M. Am. Soc. C. E., Consulting Engineer, Concord, N. H., for co-operation in research and for valuable information freely given. Others to whom the thanks of the writers are due are, Frank B. Rowell, Research Engineer, Boston and Maine Railroad Company; Lyman S. Hayes, Town Clerk, Rockingham, Vt.; and Mrs. Mary N. Bolles, Bellows Falls, Vt. Several other correspondents have contributed important data without which the history would have been less complete.

DISCUSSION

HENRY B. SEAMAN,[41] M. AM. SOC. C. E. (by letter).—This valuable paper is not merely a history of wooden bridges; it is practically a treatise on the design of wooden bridges of the more recent types. The Howe truss will be in vogue in timber countries for many years to come. The recent tendency toward iron bridges is partly the result of heavier engine loading; but it is due more especially to the increased price of timber, which would have less effect in countries where the timber is ready at hand.

Fifty years ago there were still many Howe truss bridges on the Pennsylvania Railroad. Although they were really combination bridges of wood and iron, they were classed as wooden bridges, and if iron was used unnecessarily it was considered an indignity to the wooden truss. Cast-iron angle blocks and wrought-iron tension rods were a regular part of the construction. The chord splices were of oak and were compactly bolted in place. In pro-

FIG. 29.—TIMBER HOWE TRUSS, MOUNT VERNON, N. Y., 1893.

portioning the splice, the area of the split section to that in the end thrust bearing was in a fixed ratio. Long spans (180 ft.), which still remained in service, had arch ribs on each side of the truss, which extended down below the chord to the face of the abutment, on skew-backs.

[41] Cons. Engr., Brooklyn, N. Y.

On the Erie Railroad, about ten years later, there were many Howe truss bridges of shorter span, but without the arch rib. Since the Railroad Company was avoiding the appointment of a Receiver, the bridges were used to their ultimate of safety and were badly "hog-backed" before removal. They were tested at the critical joints by the use of a small auger to ascertain the condition of the interior wood, and the holes thus made were then filled tightly with plugs. It was the saying that "a wooden bridge never falls down," which was, of course, a dangerous doctrine; but it received apparent corroboration when the Market Street Bridge, in Philadelphia, Pa., was removed, and was found to have been literally punk.

FIG. 30.—TYPICAL JACK-KNIFE DRAWBRIDGE, WESTPORT, CONN., 1905

About 1893, the writer had occasion to design and construct the overhead Howe truss bridge shown in Fig. 29, for a temporary crossing of the New York, New Haven and Hartford Railroad, at Mount Vernon, N. Y., and found no difficulty in obtaining trained carpenters to execute the work. It was of simple design, with oak angle blocks, a complete upper lateral system, and knee-braces only at the portals.

More could be written on the subject of the jack-knife draw-bridge. This is still a useful type for temporary purposes. While re-constructing the main-line bridges of the New Haven Railroad at Cos Cob and at Westport, Conn., in 1905, two such bridges, double-tracked, were constructed by the writer, and served well for main-line service during the reconstruction. One of them is shown in Fig. 30. These jack-knife draw-spans were part of the temporary by-pass trestles. For satisfactory operation of the draw-span, it is important that the gallows-frame be inflexible, and that the pile bents from which the trusses swing, be rigidly braced against movement.

JASPER O. DRAFFIN,[42] M. AM. SOC. C. E. (by letter).—Timber bridges of the type discussed in this paper are fast disappearing, particularly in Vermont where a large number have been replaced within the last ten years and especially since the flood of 1927. The writer is particularly interested in one of the bridges described, the Enoch Hale Bridge, at Bellows Falls, Vt., or the first bridge over the Connecticut River. The latest successor to this primitive bridge is a two-span, reinforced concrete arch built in 1930.

That this early bridge was considered important is shown by the account mentioned by the authors,[43] which reports that Enoch Hale "has just built" a two-span bridge, across the Connecticut River at the "Great Falls," 360 ft. long and 60 ft. above high water. It was believed to be the most elegant, the strongest, and one of the most useful bridges in America. The item went on record that while the work of building it was dangerous, the only casualties were the death of a young man who fell on the rocks in the river, and a slight injury to Colonel Hale.

The main route between Boston, Mass., and Montreal, Que., Canada, passed over a turnpike road through Keene, N. H., crossed the Connecticut River at Bellows Falls, and continued across the Green Mountains to Rutland and Burlington, Vt., and thence to Montreal by Lake Champlain. Consequently, it was an important artery of travel and transportation at that time and the Hale Bridge was a vital link in it. The reason for the selection of this place for the construction of the bridge was that the river is narrow (300 to 400 ft. wide), passing through a rocky gorge with a ledge in the middle, on which a center pier could be placed. A more favorable site for a bridge can scarcely be imagined; but not only was the location desirable from an engineering point of view, it had a considerable reputation for scenic beauty.

A question has been raised concerning the replacement of the original Hale Bridge. This arose mainly because Timothy Dwight stated[44] that, "in 1803 [the year of his first visit to Bellows Falls] the bridge erected by Colonel Hale * * * had been taken down and a new one built." He does not say that the replacement took place in 1803, but merely that it had been done by that time. In an earlier paragraph, he states that the original bridge was standing in 1797. Therefore, if his account is correct, a new bridge was built between 1797 and 1803 to replace the one built by Colonel Hale. Some doubt of the correctness of the Dwight account has been expressed, partly because there did not seem to be any local tradition of the replacement. However, there is a record that the owner of the bridge, a Mr. Guier (or Geyer), was trying to raise money by mortgaging his property at about the time of the supposed second building of the bridge. The bridge built by Colonel Hale has been described as a combination of arch and truss and

[42] Assoc. Prof., Theoretical and Applied Mechanics, Univ. of Illinois, Urbana, Ill.
[43] *The Massachusetts Spy*, February 10, 1785.
[44] "Travels in New England and New York," 1803.

neither the first bridge (see Fig. 8), nor the second one was of this type; but the description of the second bridge by Timothy Dwight would lead one to conclude that it was of the arch type, since he described it as "an obtuse arch." Taking into account all the uncertainties and the short time between the building of the first bridge and its replacement or rebuilding, it seems reasonable to conclude that the original bridge was not replaced in its entirety by a new bridge, but that it was merely closed to traffic and extensively repaired between 1797 and 1803. In the beginning of his search for information, the writer wished to believe that there was an entire replacement of the bridge, but the evidence obtained thus far does not warrant such a conclusion.

The reconstructed bridge lasted until 1840 when it was replaced by what was known locally as the "Tucker Toll Bridge," shown in Fig. 18, and which was of the covered lattice timber type with sidewalks on the outside of each truss. It was placed 15 ft. higher than the Hale Bridge which was used during the construction of the new one, after which the old bridge was cut away and allowed to fall into the river.

E. K. Morse,[45] M. Am. Soc. C. E. (by letter).—It is refreshing to read something that is not buried in a mass of intricate calculations that are known only to the author. Several facts have not been mentioned in this paper, namely, that many of the old wooden bridges, especially those constructed by Burr, were built of sized white pine, free of knots, adzed into shape, and then hand-surfaced on four sides. This was especially true in the Burr Arch Bridge across the Allegheny River at Ninth Street, in Pittsburgh, Pa. (which the writer contracted to remove in 1890-91), and also in a bridge of the same design built by Burr at Saltsburg, Pa., across the Kiskiminetas River.

In inspecting the bridge at Saltsburg some years ago, the writer found the name of one of the carpenters who evidently had planed the beautifully shaped and finished members of the arches. The man had engraved his name in the top of one of the arches which was just as bright and as clear as the year it was erected. All the timbers going into these bridges were seasoned from two to four years before being placed in position, which accounts for the accurate work and the condition of the bridges after years of heavy travel. On the Ninth Street Bridge there was not an open joint on any of the spans, averaging from 178 to 210 ft, center to center of skew-backs. The white oak flooring, like the white pine, was seasoned in the old-fashioned way and surfaced on all four sides. Teams and wagons passing over the floor moved so smoothly that the edges were not frayed. In practically all the bridges that were erected by Burr, the floors were originally laid dust-proof and because they were not exposed to the elements, they remained intact for many years.

In taking down the old Ninth Street Bridge, heavy vehicle and foot travel had to be maintained and, in order to facilitate erection, the roof was removed, the trusses were cut to the line of the arch, and 1½-in. bolts were placed through the edge, and the verticals supporting the floor system clamped tightly to the arches wherever they showed signs of weakness. In

[45] Engr. Member, Water and Power Resources Board of Pennsylvania, Pittsburgh, Pa.

order to prevent excessive vibration, a two-post tower with sway-bracing was erected at the center of the arches. There was considerable vibration in the arches when they were cut loose from the floor system and the tower was removed, but when $1\frac{1}{2}$-in. manila rope was connected to the top of each arch and it was attempted to pull them over and dump them into the river, they yielded readily for about 3 or 4 ft and then resisted even the 5 tons of horizontal pull that was put on them, and eventually had to be removed panel by panel.

It was found that the members in all the arches had completely lost their fiber to the extent that it was almost impossible to cut one of these timber arches with an axe. It was as difficult to lift a chip from one of these white pine timbers as it is with a red gum or green sycamore. The timbers were bright. During the erection of the new bridge, they supported a load three or four times as great as that for which they were designed and, in no case, did any member of the arches show any sign of weakness before, during, or after being taken down. A match factory in Chicago, Ill., had bought the timber in the white pine arches, and needless to say the contract was cancelled when it was discovered that white pine which is so easy to chop and so beautiful to work, had become as tough as gum.

JOHN W. STORRS,[46] M. AM. SOC. C. E. (by letter).—This paper is not only interesting and informative, but instructive, and must have required much thought, research, and investigation.

Old wooden bridges have recently been the subject of a number of brochures. Many people are showing an interest in these structures. As an exemplification, there was an unique service at the site of an 82-year old bridge in the Town of Boscawen, N. H., on the afternoon of December 3, 1932. This is a covered bridge of the Burr type with arches, and it spans the Contoocook River. It has been in constant use since 1850. Repairs have been completed recently, and it will do service for a long term of years to come.

The bridge at Woodsville (in the Town of Haverhill, N. H.) built in 1828, and the one in Bath Village built in 1832, are still in service and apparently taking care of the traffic in a satisfactory manner.

The writer would be one of the last to cast any reflections, to find fault with or to criticize the earlier American builders of bridges. They were limited as to means and material; had little to go by except trial and experience. Some of the so-called hybrid or mongrel bridges did, or have done, extraordinary and remarkable service, and the constructors and bridge carpenters of those days "did the best they could; angels could do no more."

C. J. HOGUE,[47] M. AM. SOC. C. E. (by letter).—This paper is one of great historical and informative interest and value, at least to one connected with the timber industry, and the writer hopes that it may be to others, as well.

---

[46] Cons. Engr., Concord, N. H.

[47] Engr., West Coast Lumbermen's Assoc., Inc., Seattle, Wash.

It brings the history and practice in wood bridges to date, and the writer believes that the present is the eve of a fourth era in wood bridge design and construction, which will be of even greater interest and importance, from both engineering and architectural viewpoints, than the three recorded in the paper.

Temporarily, bridge design and construction in wood have been somewhat eclipsed by other materials. These materials have been new; wood construction has been old. They have been replete with new technical studies and developments; wood technology, old as it is, has lacked definite information on some points.

To-day, as not for many years, the bridge designer is faced with the necessity for the utmost economy and, probably as never before, with the question of the length of useful life and early obsolescence. A timber supply is amply available, recent research has established authoritative information on most points in question, new methods of framing have been developed, and treated timber offers material of economic permanence, ease of remodeling, low cost of removal, and salvage value on removal.

During recent years, the Forest Products Laboratory, of the U. S. Forest Service, at Madison, Wis., has published the results of technical studies which have taken the mystery and variance from the calculation of the bearing strength of wood posts, of bearing on wood at various angles to the grain, of the bearing strength of bolts on wood, of the calculation of shear in wood beams and stringers, and of the basic principles of calculating the strength of laminated wood beams and posts. The ring dowel connector, developed in Germany during recent years, opens a vast new field of economy and flexibility in wood framing.

It has been shown by experience that the durable life of treated timber may be assumed safely at forty years or more; it seems to be the consensus of opinion of bridge engineers that the useful life of a bridge in these times of rapid change may not be assumed safely at more than twenty to twenty-five years, often even less, due to early change in location.

The public has been taught to expect beauty as well as utility in its bridge structures. During 1932 a number of bridges of appealing architectural character, with little increase in cost, have been built in Oregon.[48] Such framing connectors as the ring dowel open the field for wood bridge construction to long-span arches and arched trusses of imposing size and appearance. The writer believes that a new chapter in wood bridge construction is about to be written.

William A. Oliver,[49] Assoc. M. Am. Soc. C. E. (by letter).—The reading of this paper has been a pleasure and an inspiration. The education of an engineer is not complete until he has become acquainted with the work of the giants who have gone before him. It is through such efforts as the one under discussion that this part of an engineering training is obtained.

---

[48] *Civil Engineering*, September, 1932, Vol. 2, No. 9, p. 549.

[49] Urbana, Ill.

In their "Synopsis," the authors write, "but it will appear that, where timber is plentiful * * * wooden bridges still have 'a reason to be.'" Later, they state (see "Stringer Bridges"), "although it can be said that the day of wooden bridges has passed, it is a fact that many hundreds of miles of wooden trestles, stringer, and truss bridges are still in service in the United States." It is not entirely clear just what the authors' conclusion is regarding the present adaptability and economy of timber bridges. The writer would like to discuss the paper briefly from this point of view.

One of the important objections to timber bridges and structures generally has been the so-called "lack of permanency." The writer does not believe that the facts justify this idea. By proper maintenance a timber bridge may be just as permanent as a bridge built of some other material. For centuries, and by consistent and precise maintenance, the Japanese have made their timber bridges perpetual. Numerous examples of timber bridges several decades old can be found in the United States, and many of these bridges have had only ordinary care during these years of constant use. Certainly fifty years is a long life for any ordinary bridge of any material in the United States. Some examples of present-day wooden bridges are briefly discussed in the following paragraphs.

FIG. 31.—SELBY BRIDGE OVER THE RIVER OUSE IN YORKSHIRE, ENGLAND.

According to Mr. E. Jervoise,[50] Selby Bridge over the River Ouse, in Yorkshire, is probably the oldest timber bridge in England. It was built in 1790 and is still in use (1933). The bridge has a swing section which, in Fig. 31, is shown thrown from its pivot. Mr. Jervoise states that through this mishap the bridge was out of use for some time.

[50] Ancient Bridges of England Series, Architectural Press, Westminster, S. W., London, England.

Of course, modern examples of wooden bridges are plentiful on the West Coast of the United States, but, in 1932, the State of Ohio found it expedient to build ninety-three timber highway bridges. Most of these structures were of a simple trestle type, built of timber produced in the State. All material was impregnated with preservative, all framing being done in advance of treatment. It is interesting to know that a mid-Western State had timber in this quantity available for use, and that one of the more progressive States did find it economical to adopt such an elaborate program of timber bridge construction.

One of the earliest of materials of construction, timber has been developed for structural purposes by "rule-of-thumb" methods. Due to the inertia that these "rule-of-thumb" methods obtained through the years, it has been difficult to make the ordinary timber user appreciate the fact that greater economy as well as strength could be obtained through a more scientific use of the material. However, through the efforts of the various lumber manufacturers' associations and the work of the materials testing laboratories, this difficulty is gradually being overcome and the general trade is evincing more and more interest in researches in the use of timber. Many examples of present-day research could be given as examples of what is being done along these lines. As far as the testing laboratory is concerned timber is almost an ideal material, conforming beautifully with the accepted theories of mechanics.

Referring to the work of Messrs. Brangwen and Sparrow,[14] the authors state in their discussion of Roman bridges, that very little is known of the construction of Pons Sublicius, "hallowed by traditions." A longitudinal elevation and a cross-section of this bridge has been published[51] by Mr. H. G. Tyrell. The writer wonders whether the authors question the authenticity of this illustration.

PHILIP G. LAURSON,[52] M. AM. SOC. C. E. (by letter).—The authors have contributed a paper of great value to civil engineers who are interested in bridges. Much of the information is not available elsewhere.

It is natural and proper to emphasize the American developments as the authors have done. It is true, however, that wooden bridges were used in Europe a century ago to a greater extent than is commonly realized.

The *Minutes of Proceedings* of the Institution of Civil Engineers of Great Britain, and other European engineering publications, contain numerous descriptions of wooden railway and highway bridges. In 1846, John Rennie, in his Presidential Address, reviewed the developments in all branches of Civil Engineering, including wooden bridges. He reported at that time[53] that the trussed system had been applied with considerable success in some well-constructed bridges across the Tyne River and in several other places. The system of combining small curved pieces of timber connected in the form of an arch, adapted for large spans, was first introduced by C. F. Wiebeking in 1826, according to Rennie, when a bridge of 100-ft span was constructed on the

[51] "History of Bridge Engineering," p. 23.

[52] Assoc. Prof. of Eng. Mechanics, Sheffield Scientific School, Yale Univ., New Haven, Conn.

[53] *Minutes of Proceedings*, Inst. C. E., Vol. V (1846), p. 33.

Ancholme with complete success. He reported also that the lattice bridge had recently been introduced on the Birmingham and Gloucester Railway and on the Dublin and Drogheda Railway and that it had been found economical and simple in construction. A number of European types of wooden bridges, with details, was described by Benjamin Green in the same year[54] in a paper entitled "Arched Timber Viaducts."

Isambard Kingdom Brunel designed many timber railway bridges in a great variety of types.[55] Brunel was particularly concerned with the prevention of decay and, as early as 1835, was in communication with Faraday as to the best method of testing the extent to which Kyanizing solution penetrated into wood. It is reported that Brunel made a careful study of all the different methods of preserving timber, and used the more successful of them on a very considerable scale.

Many of the European engineers of these early times were well-educated men of high technical skill. Carl Friedrich von Wiebeking (1762-1842), mentioned by Rennie, was Director-General of Roads and Bridges in Bavaria, and was the author of several treatises on hydraulics, mechanics, and structures. He worked out proportions for his laminated wood arches for spans as long as 600 ft. Bridges of this type were built in England and on the Continent in considerable numbers. The authors' Fig. 6 is of this type, but in most of the bridges the bent planks were thinner — 3 in. being a common thickness. As many as sixteen or twenty thicknesses of bent planks were bolted together to form the arch rib.

Another good source of information concerning European timber bridges is the set of publications on the "Theory, Practice, and Architecture of Bridges," edited and published by John Weale about 1843; it contains two volumes of fine plates. The early volumes of *L'Illustration* also contain pictures of wooden bridges in the days when railroads were still objects of curiosity.

It is true, of course, that because the use of timber was less advantageous in much of Western Europe, than it was in the United States, bridges of other materials have always predominated.

The practice of assembling trusses on land before erection must have been more common than the authors imply. David Stevenson, a Scottish civil engineer, made an extensive tour of the Eastern United States and Canada, about 1837, for the sole purpose of inspecting engineering works.[56] He reported seeing many Town truss bridges on the Philadelphia and Reading, and on other, railroads.

He also noticed the Long bridge in many parts of the country, the best specimens being on some of the railways in the neighborhood of Boston, Mass. His account states, furthermore, that:

"The timbers of which Town's and Long's bridges are composed, are fitted together on the ground previous to their erection on the piers. They are again taken asunder, and each beam is put up separately in the place which it is to occupy, by means of a scaffolding or centering of timber."

[54] *Minutes of Proceedings*, Inst. C. E., Vol. V (1846), p. 219.
[55] "Life of I. K. Brunel, Civil Engineer," by Brunel, Lond., 1870.
[56] "Sketch of the Civil Engineering of North America," Lond., 1838.

RICHARD S. KIRBY,[57] M. AM. SOC. C. E. (by letter).—The authors of this paper have covered in a thorough and interesting fashion a field too long neglected. The writer was especially pleased with their emphasis on personalities of the period. In this connection he ventures to add a word concerning the McCalls Ferry Bridge over the Susquehanna River, the work of Theodore Burr, Connecticut-born engineer and contractor, of Harrisburg, Pa. This bridge was one of Burr's great achievements, but little is heard of it because of the fact that it was carried down by the ice in 1817 after only two years' service. The writer believes that it had the longest span of any wooden bridge ever attempted in America, and that only that of the Schaffhausen Bridge had exceeded it.

In a long letter written by him shortly after he finished this bridge, to his friend, Reuben Field, bridge builder of Waterford, N. Y., Burr rehearsed graphically his thrilling experiences in erecting the bridge during the winter of 1814-15. The arches were fabricated in halves and raised at first on bents of varying heights supported by floats ranged along and parallel to the shore nearly $\frac{1}{4}$ mile below the site. As the ice increased in thickness, Burr decided to use it instead of the floats. He finally slid the entire structure up stream on rollers. Man-power was abundant in those days, in that locality at least, for Burr commends the inhabitants for their willing help, which was perhaps gratuitous.[58]

WELLS N. THOMPSON,[59] ASSOC. M. AM. SOC. C. E. (by letter).—In the early stages of the development of wooden bridges, the crib or corbel type was one of the most interesting. Examples of this type of bridge are fast disappearing as they are replaced by more modern structures, but there are a few remaining. Fig. 32 shows a very interesting one in the Far East.

This old bridge is well known throughout Japan, and it is shown or mentioned in many native wood-block prints, poems, and songs. It is located in Central Japan, Yamanashi Province, and supplies its own name to the nearest town, "Saruhashi" or "Monkey Bridge." It is fifty miles west of Tokyo and spans a mountain stream at a point where it is approximately 80 ft wide between abutments. Age is evident in the entire structure, but repairs and replacements have been made as required so that it continues in good condition and in general use for the traffic in that district, which consists largely of pedestrians, jinrikishas, and ox-carts. The roadway is about 12 or 14 ft wide.

The Japanese excel in timber construction, and are very careful about protecting their structures against rot and decay. In this bridge, as in all other structures where framing is exposed to the weather, they have protected each member in the supporting cribs with an individual roof or watershed. Such care has preserved many of their very old and notable structures.

---

[57] Assoc. Prof., Eng. Drawing, Sheffield Scientific School, Yale Univ., New Haven, Conn.

[58] The letter appeared in *Niles Weekly Register* (Baltimore, Md.), November 18, 1815. It is copied in the Lancaster Historical Papers, Vol. XI, p. 133.

[59] Danbury, Conn.

The abutments are field stone masonry, built against the rock walls, and have probably been added recently, in rebuilding or reinforcing the old stonework, perhaps after damage by earthquake.

FIG. 32.—OLD CORBEL TYPE OF BRIDGE, YAMANASHI PROVINCE, JAPAN.

BENJAMIN WILDER GUPPY,[60] M. AM. SOC. C. E. (by letter).—In connection with Fig. 28 of the paper, the authors point out the important warning that the bridge builder cannot be too careful in designing the bolted and spliced joints of a timber bridge. A few examples should serve to emphasize this importance.

A draftsman who has had experience only in detailing metal structures has considerable difficulty when he first attempts to design a timber structure. In such a case the writer has told the man that if he designs a steel bridge while standing on his feet, he should stand on his head while designing a timber structure. In other words, the processes are reversed. In steel designing, the sections of the main members are determined first and then the details designed to transmit the stress properly. In straight timber work it is the design of the connection that generally determines the dimensions of the main members. In metal work the unit stresses to be considered are those of tension, compression, shear, and bearing. In timber the unit stresses to be accounted for are tension, compression, bearing on end grain, bearing across grain, shear parallel to fiber, and shear normal to fiber on planes parallel to and normal to direction of fiber.

[60] Engr. of Structures, B. & M. R. R., Boston, Mass.

The simplest case is a post in a trestle bent. In a steel structure the area of the post is determined by dividing the stress it is to carry by the allowable unit stress determined by some approved column formula. The dimensions of the timber post are determined by the stress in the post divided by the allowable stress in bearing across the grain on the cap or sill. A moment of reflection will show that it will be a very long slender post to require an allowable unit compressive stress as low as the unit bearing stress on side grain of timber.

The next step in advance is a king-post truss and to illustrate his statement the writer presents a design using timber entirely, use of iron being restricted merely to small bolts to hold the various pieces together (see Fig. 33). The king-post is in tension. To determine its proper dimensions both upper and lower connections must be considered. At the lower connection the size of the post is determined by the net area required in tension, plus the area required to transmit the stress by bearing across the grain. At the upper connection the area of the post is determined by the net area required to carry the tension, plus the area required to transmit the stress by bearing on the end grain. The larger of these areas governs. At the upper joint the strut must be framed both to give sufficient area for the vertical component in the strut at the unit stress for bearing on the end fiber and the horizontal component in the strut at the unit stress for bearing across the grain. The

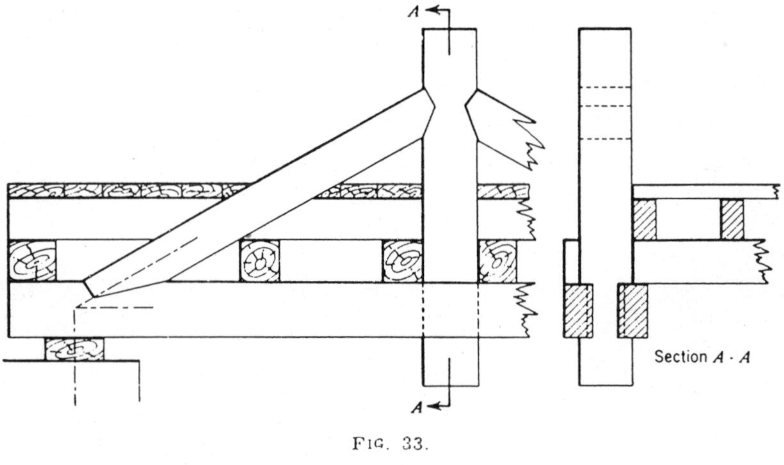

FIG. 33.

distance from the daps to the end of the king-post must be sufficient to prevent the timber from shearing beyond the daps. At the bottom of the strut the connection is reversed from that at the upper end as the horizontal component must be carried into the bottom chord through the end bearing and the vertical component through the side bearing. The bottom chord must extend far enough beyond the dap to give sufficient strength in the shear along the grain. The size of the bottom chord at the end joint is determined by the depth of dap necessary to transmit horizontal stress from the strut and the depth of stick required below the dap to carry the tension in the bottom chord plus any bending that may be induced by eccentric connections or methods of loading.

When it is necessary to splice a tension member the design of the splice is necessary before sizes of timbers can be determined. This is not submitted as an example of good modern practice, but to illustrate principles of design in timber and to show what will have to be done if metal for details is not available. The use of metal angle blocks to distribute stress at connections will permit of detailing compression members in accordance with an approved column formula and the use of rods for tension members and joint bolts will greatly simplify detailing connections. Figs. 16 and 28 in the paper by Messrs. Fletcher and Snow show how efficiency of tension splices can be increased by a free use of metal.

IVAN E. HOUK,[61] M. AM. SOC. C. E. (by letter).—The authors have contributed an extremely interesting and valuable account of the history, design, construction, and life of wooden bridges.

FIG. 34.—VIEW OF TIMBER TRUSS COVERED HIGHWAY BRIDGE OVER THE STILLWATER RIVER IN SOUTHWESTERN OHIO.

In connection with their discussion of retrieving weak and injured bridges, an account of the rather novel repairs shown in Fig. 34 may be worth while. This bridge is one of the old timber truss, covered highway bridges, still so common in the Eastern United States. It spans the Stillwater River, of

[61] Senior Engr., U. S. Bureau of Reclamation, Denver, Colo.

Southwestern Ohio, about two miles above the Englewood Dam. It withstood, safely, the disastrous flood of March, 1913, as well as all the previous floods that had occurred after it was built. When it was built, or whether it is still in existence, is not known. However, it must have been about fifty years old at the time the Englewood Dam was completed. The Englewood Dam is one of the five large flood-prevention dams built by The Miami Conservancy District about 1920 or 1921.

Prior to the completion of the flood-prevention works the bridge was located on the National Road and carried all through traffic along U. S. Highway 40 in that section of the State. Since that time through traffic has been routed over the top of the dam. However, the bridge was still in good condition at the time the dam was built, and it was desired to keep it in operation after the through traffic was diverted. Consequently, it was necessary to anchor it in place, to keep it from floating away during flood periods when the back-water above the dam might rise 30 or 40 ft above the roadway surface. In other words, the construction of the dam below made it necessary to hold the bridge down as well as up.

Iron rods, $1\frac{1}{4}$ in. in diameter, with an eye at the bottom and threads at the top, were placed vertically between the timbers of each truss, a rod being placed on each side of the pier and at each abutment. The upper end of the rod passes through the heavy timber which constitutes the upper chord and is held in place by a large iron washer and nut. The eye at the lower end is tied to the foundation by four steel cables, attached to iron bolts set in the masonry about 1 ft apart and about 12 ft below the lower chord.

The old, timber truss, covered bridges in the Miami Valley went through the destructive flood of March, 1913, as well as, if not better than, those of more modern design. Of course, this was because the masonry foundations had been either founded on rock or carried down to a sufficient depth below the bed of the stream, and had then been built to a height well above the maximum previous high-water level before placing the superstructure.

WILLIAM H. BAKER, JR.,[62] ASSOC. M. AM. SOC. C. E. (by letter).—The various developments in the use of wood for the construction of bridges, is illustrated clearly in this paper. It covers the development from the most simple type, through all the stages of design, to the present standard. The writer believes that the development of the wooden bridge has been a great influence on the history of the world, both in military conquest and social growth through the medium of communication. The wooden bridge is still used extensively in the United States, despite the great progress in the use of steel and reinforced concrete in the last twenty years.

The wooden bridge is still used much more extensively than bridges of steel and reinforced concrete in Los Angeles County, California. There are approximately six hundred bridges of wood construction in use at the present time. They consist principally of **A**-frames up to 42-ft span, together with pony Howe trusses for longer spans. The reason for this extensive local use

[62] Cons. and Structural Engr., Long Beach, Calif.

of the wooden bridge is due to the inadequate finances available to construct a more permanent type of crossing over the many dry washes in the county—most of them dry, except during flood stages—and also because as yet flood-control channels have not been established in many instances. Some of these wooden bridges over dry washes are as long as 800 to 1 000 ft, and to have built them of steel or concrete would have increased the cost many times that of the wooden structures; and after the flood-control channel had been established definitely, the channel would perhaps be much narrower than the original bridge span.

Since about 1925 or 1926 it has been necessary to widen many existing highways in Los Angeles County. The early bridges had 20-ft roadways—inadequate for present traffic needs. Most of the bridges now constructed have a roadway width which will be sufficient to provide for present traffic demands and for future widening of the highway. It was necessary, therefore, that the County Road Department adopt a policy in which all future bridges would provide roadways of at least 30 ft. The wooden bridges, both the A-frame and the Howe truss type, were constructed using large Oregon pine floor-beam members. These floor-beams for a bridge without skew were 24 ft long for a 20-ft roadway, being made up of two 12 by 36-in. timbers bolted together. For skew bridges, the floor-beams were longer, as the angle of skew increased, making these beams from 24 ft to approximately 36 ft in length. It was very difficult to build wooden bridges for a 30-ft roadway utilizing wooden floor-beams, due to the length requirement of such a large wooden section. At the same time, the cross-sectional area required for the wooden floor-beams greatly reduced the head-room and opening for the waterway.

At the time that the Carnegie beam sections using flat flanges were introduced, the Bridge Department of Los Angeles County designed and developed a wooden truss bridge of the Howe type for 30-ft roadways, and for various angles of skew, utilizing the new Carnegie beam sections of 30, 36, and 42-in. depth for floor-beams replacing the old wooden type beams. A very simple method of constructing a proper connection for the hanger rods and extension brackets for the pony truss braces was designed and developed. This use of the Carnegie beam section as floor-beams, which has now (1933) been standard for five years, permitted the design and construction of wooden bridges of 30-ft roadway and a much greater angle of skew than was possible with wooden floor-beams. At a later date, about two years after steel floor-beams were utilized, the lower tension chord member was designed, with steel channels to replace the old, built-up, wooden type lower chord.

The principal difficulty experienced with bridges of the wooden type was the dry rot that developed in the end-posts and flooring. Several types of construction have been tried over a period of years to eliminate this, including the use of 2 by 6-in. timbers of Oregon creosoted pine, laid on edge and the utilization of diagonal planking overlaid with transverse planking at right angles and surfaced with asphalt or macadam.

It is impossible to state or give true results at the moment as to which type or method of wooden floor construction is superior. Later, due to the difficulty in maintaining the wooden type of floor construction, a reinforced concrete floor-slab system placed on timber stringers was tried as an experiment. Several bridges utilizing such construction have been in place for about four years, and to the present time (1933), are in good condition. However, it is problematical whether this method will prove to be entirely successful and will replace the old method of wooden floor construction. The bridges that the writer last described are not strictly wooden, but are a combination of wood, steel, and concrete.

The use of the wooden bridge in Los Angeles County will probably continue for some years to come, until flood-control channels are definitely established and until new means of financing permanent structures of concrete and steel are advanced. The use of the wooden bridge is very costly, due to the heavy maintenance charges involved. However, they will have to serve until something more economical from the standpoint of first cost is substituted. This paper is of unusual interest and is a great contribution to civil engineering history.

OREN REED,[63] ASSOC. M. AM. SOC. C. E. (by letter).—The authors are to be commended for their excellent history of the development of different types of wooden bridges, a class of engineering structure that is fast disappearing in Central and Eastern United States. Since the structures described were located in the Eastern States, the writer wishes to refer to a few wooden bridges of Indiana to show the wide range in the use of such bridges. Most of these old structures, which are still in use, are on little back roads, off the main highways, and in out-of-the-way places; but a few are on major routes, such as the multiple-span bridge on the present U. S. Route No. 50 over the East Fork of White River, near Medora, and the two-span bridge on State Road No. 52, at the edge of Rushville, over Flat Rock River.

Nearly all the early covered bridges in Indiana were of the Burr truss type. The Howe truss soon superseded the Burr truss, to a large extent, in the East, but was not commonly used in Indiana, and probably not at all before 1865. The National Road, the Michigan Road, and the Vincennes-New Albany Pike were the early important roads and most probably the ones on which the first permanent bridges were built. For most of their lengths, these have remained important roads, and only a few of the original bridges remain.

Mr. J. J. Daniels, of Rockville, and the Kennedy family, of Rushville, are largely responsible for the great number of excellent timber bridges in the State, many of which are still in use. Mr. Daniels built about fifty bridges, mostly in Parke and Vermillion Counties. He built a bridge over Sugar Creek, in Parke County, in 1861, which is still in use. This structure has a clear span of 200 ft and is the longest single-span covered bridge of record in Indiana. The Kennedy family built about sixty covered bridges

[63] Constr. Engr., Emergency Conservation Work, Camp 2, Lincoln City, Ind.

in all, the first being at Dunlapsville, in Union County, in 1870, and the last near Fountain City, in 1918. Many of these bridges are still in use in Southeastern Indiana. The Burr truss bridge, at Rushville, built by the Kennedys in 1883, has two 125-ft spans with a 16-ft clear roadway, and is unusual in that it has an arcaded sidewalk on each side.

Two of the oldest timber trusses, still in service, are the Ramp and Raccoon Creek Bridges formerly on State Road No. 43, near Greencastle, which were built in 1837 and 1838. These bridges were constructed of local material, except for a few hand-made iron bolts and brace-pins. Both have two 11-ft roadways carried by three trusses, one of which separates the two roadways. The Ramp Creek Bridge has a clear span of 91.5 ft and the Raccoon Creek Bridge, one of 120 ft.

The Ramp Creek Bridge was removed in 1932 and re-erected over Salt Creek, at the north entrance to Brown County State Park, by the Indiana State Highway Commission. The arch ring, top chords, rafters, and roof framing are of yellow poplar; the posts, lower chords, and floor are of oak; while the floor-beams, joists, and sub-flooring are of walnut. Members that had decayed to any extent, were replaced by new timber. Oak was substituted for the damaged walnut. In all, 10 to 15% of the members were replaced. The abutments, which were carried to rock, are of concrete, faced with sandstone masonry. This timber bridge will have ample strength to carry the park traffic and should stand for many years as a monument to the early bridge engineers and their ingenuity in the use of local materials. Raccoon Creek Bridge is also to be replaced by a modern bridge to care for the heavy through traffic on State Road No. 43. Another timber bridge, maintained by the Indiana Department of Conservation, spans Sugar Creek, in Turkey Run State Park, at the Narrows Crossing. It has a span of about 100 ft.

R. P. DAVIS[64] and L. L. JEMISON,[65] MEMBERS, AM. SOC. C. E. (by letter).—The authors of this paper merit the thanks of the Engineering Profession for their effort to record the history—from an engineering standpoint—of the fast-disappearing wooden bridge. Much of a general nature has been written on this subject in the last few years, most of which is, however, not from the engineer's point of view.

During the latter part of the Eighteenth Century, in building highways to connect the Ohio country with the Eastern seaboard, the stone arch was widely used, as exemplified by those on the National Highway between Hagerstown, Md., and Wheeling, W. Va. A little later the wooden bridge became the more popular, and from the beginning of the Nineteenth Century until after the Civil War wooden bridges were very common west of the Allegheny Mountains, the Burr type being almost exclusively used, without arches for the shorter spans. Nearly all these bridges were protected with roofs and side covering, with an open space at the top to permit proper ventilation.

[64] Dean, Coll. of Eng., West Virginia Univ.; Cons. Bridge Engr., West Virginia State Road Comm., Morgantown, W. Va.

[65] Bridge Engr., West Virginia State Road Comm., South Charleston, W. Va.

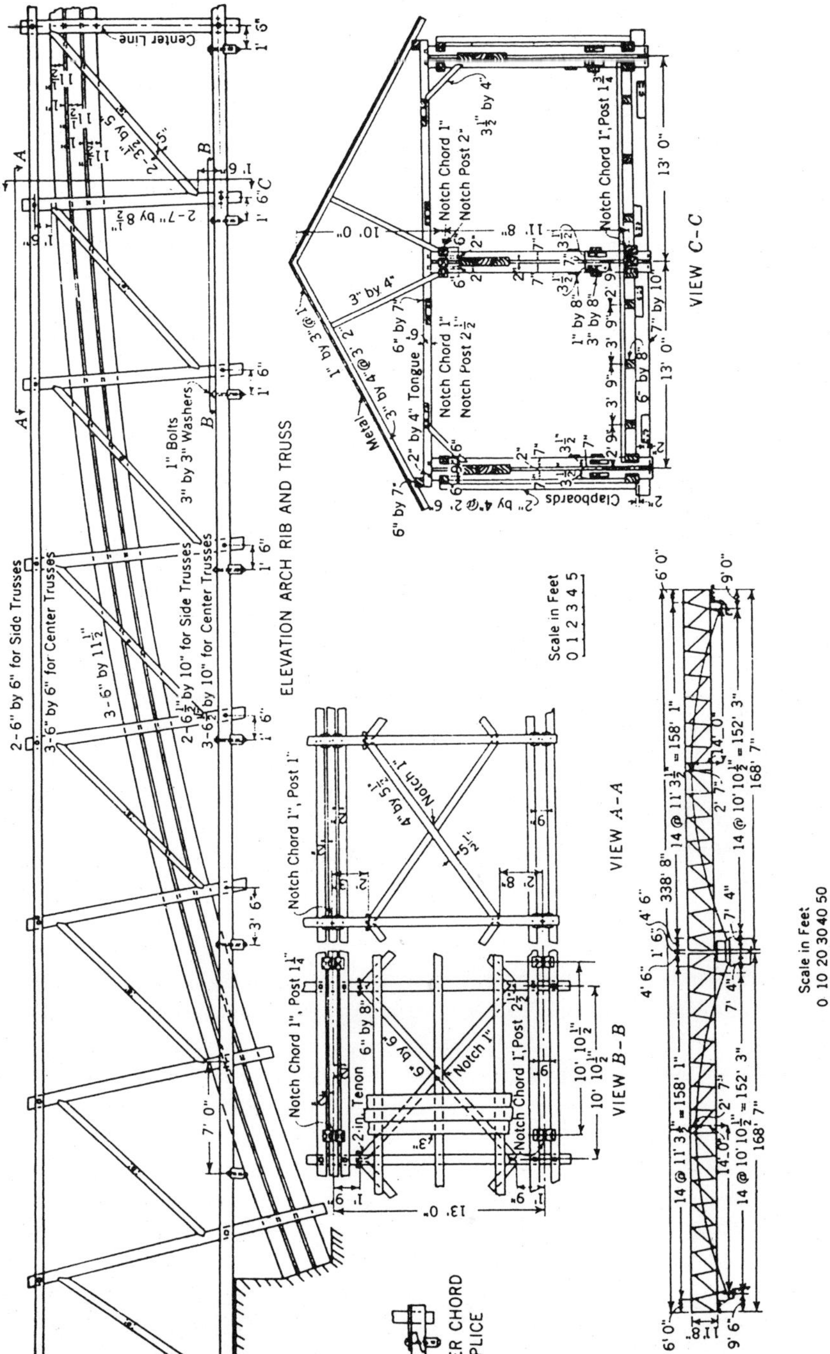

FIG. 35.—BRIDGE OVER CHEAT RIVER, NORTHWESTERN TURNPIKE, PRESTON COUNTY, WEST VIRGINIA

Of particular historical value is the two-span combination arch and truss bridge over the Cheat River above Rowlesburg, W. Va. (see Fig. 35). This structure is on the old Northwestern Turnpike (later known as Federal Route 50), a highway built in 1832 under Col. Claudius Crozet, who had been an officer of artillery under Bonaparte in the Russian campaigns.

Completed in 1834 it has given excellent service for a period of more than 98 years. The material is white pine with all timbers hewn by hand; all bolts are of wrought iron with large heads. The bridge was built at a cost of $18 000, or $2.40 per sq ft of roadway, on the basis of a total length of 339 ft, and a clear roadway width of 22 ft. The arch ribs are composed of three 6 by 11½-in. timbers, placed vertically one above the other, an arrangement not as common as placing them side by side horizontally. However, by this arrangement (see Fig. 35), less notching is required, and the full section of the arch may be secured. The stringers, 6 by 8 in. in section and spaced 3 ft 9 in. apart, are larger than those customarily found in highway bridges of the early days. The 7 by 10-in. floor-beams are hung from the lower chords by two 1-in. bolts. Why the more common and simpler method of placing these beams above the chords was not used, is not apparent. However, this method of support gave complete satisfaction, because the original bolts are still (1933) in an excellent state of preservation.

It is interesting to note that the tension web members of the trusses are not placed vertically, as is more usual, but inclined in a direction approximately normal to the arch ribs. Loads from the floor-beams first come to the trusses and are transferred to the arch rib by notching the so-called verticals and by placing wedges between the three rows of timbers forming the arch rib. It will be observed that these trusses have only a single system of diagonals. As these diagonals are framed so that they take only compression, the question naturally arises as to why a double system was not used. The sticks composing the arch rib and the lower chords are three panels in length. The type of splice used for the lower chord would be considered now a very poor one for a tension connection.

During recent years a double floor has been used on this bridge, together with longitudinal runways. These were necessary on account of the wide spacing of the stringers. When heavy truck loading was introduced, considerable weakness developed in the floor-beams in resisting horizontal shear near the ends. This weakening was due to the notching of the floor-beams at their ends and also to the cutting of holes to permit framing the diagonals of the lower lateral system.

Most of the timbers composing the trusses and the arches are in an excellent state of preservation, except the arch ribs near the springing points where, due to failure to carry the siding down in the early years, considerable rotting developed. The bridge has also suffered under the action of some severe storms, as a consequence of which the trusses and arches are badly bowed out of line. The substructure is composed of stone masonry laid with close joints, without mortar. It is in good condition.

The bridge illustrated in Fig. 36 was on the famous James River and Kanawha Turnpike which was opened to traffic in 1800. The bridge was built in 1872 across the Greenbrier River, at Caldwell, W. Va. It had two spans, the arches being 228 ft long. Like the Cheat River Bridge it had a double roadway, the outside trusses being 25 ft, center to center. It was the second bridge to be built at the same location; the first one, constructed in 1807, was burned during the Civil War. The second bridge was built at a cost of $12 500, or $1.40 per sq ft of roadway, the superstructure resting on the original pier and abutments. When this masonry was torn down in 1932, the piers and abutments were found to be composed of stone masonry shells backed with small stones.

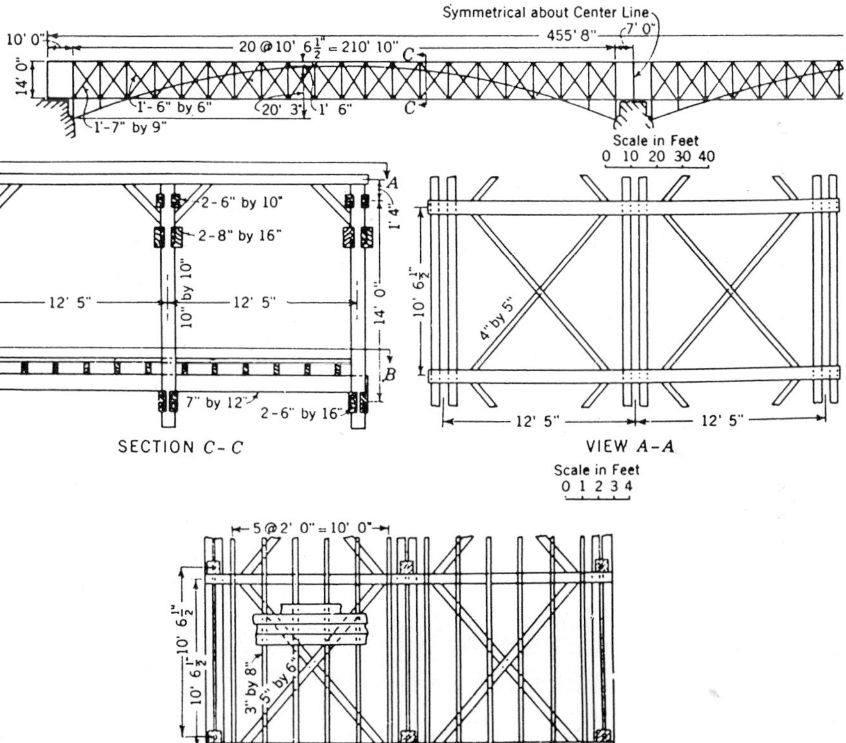

FIG. 36.—BRIDGE OVER GREENBRIER RIVER, AT CALDWELL, W. VA.

The floor system has 3 by 8-in. stringers spaced on 2-ft centers, resting on 7 by 12-in. floor-beams, which, in turn, rest on the lower chords of the trusses. As the trusses and arches resemble very closely those to be described hereafter for the Philippi Bridge, neither these details nor the roof covering is shown on the drawing.

The bridge illustrated in Fig. 37 is over the Tygarts River, at Philippi, W. Va. Built in 1852, it is said to have been the scene of the first land

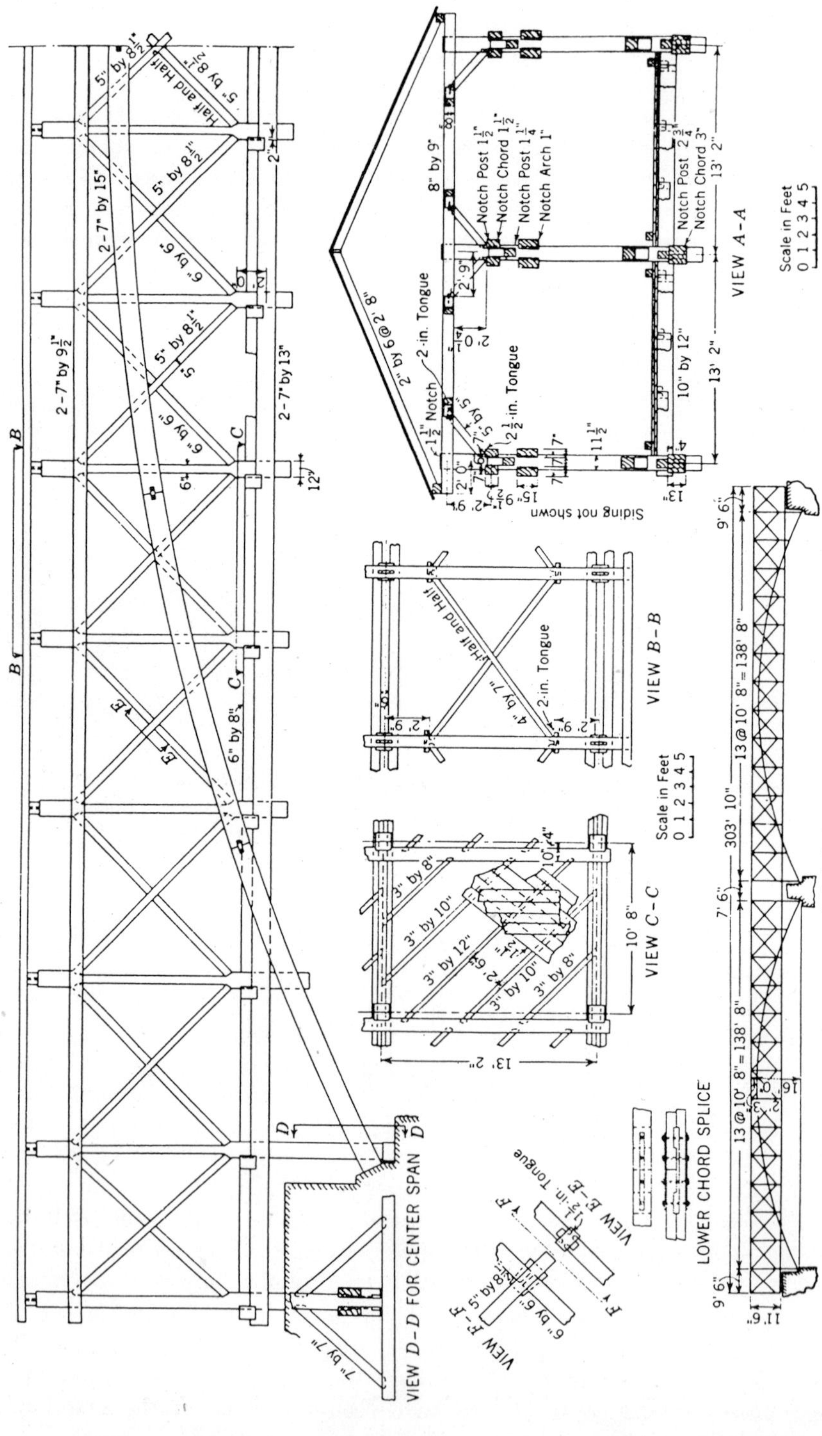

Fig. 37.—Bridge Over Tygarts River, at Philippi, W. Va.

battle of the Civil War. It has two spans, the arches being 138⅔ ft long. The bridge has a peculiar stringer system composed of diagonal 3 by 8-in., 3 by 10-in., and 3 by 12-in. beams, notched half and half into the floor-beams and into 6 by 8-in. timbers placed on, and running parallel with, the lower chords. This type of connection has proved very weak under concentrated loading and has given considerable trouble by horizontal shearing at the ends of the stringers. No lower lateral system was used on this structure, probably for the reason that the stringers, being framed into the floor-beams and pinned as well, were expected to serve as bracing. The excellent type of fish-plate joint used for lower chord splicing should be noted.

FIG. 38.—INTERIOR VIEW OF PHILIPPI BRIDGE

The bridge was built by Lemuel Chenoweth, of Beverly, Va. (now W. Va.). Mr. Chenoweth was a bridge architect by profession, and it is said that his designs were original with him, and every principle was worked out with mathematical accuracy. He knew beforehand the shape and size of every piece of timber used in the framework of his bridges. The contract for this bridge was let at Richmond, Va. Bidders were present in large numbers with all kinds of models and plans. As far as appearances went, it is said that some of the New England Yankees had models of perfect form and beauty, painted and enameled in the highest art. Mr. Chenoweth's plain wooden model attracted little attention until he placed it on two chairs, one end resting on each, and then stood on his little bridge, and called on the other architects to put theirs to the test by doing the same. This feat got him the job of building the bridge.

Considerable decay of the arches has occurred near the springing lines, but all timbers that have been properly protected are in good condition. After eighty years of continuous use this structure still lines up almost perfectly.

The details of the trusses and arches are quite similar to those of the Waterford Bridge (Fig. 15) described by the authors. Fig. 38 gives a view of the interior of the Philippi Bridge.

The only metal in any of the bridges described was in the form of bolts, which were freely used to hold the various members, and their segments together; but in no case, except the floor-beam connections and the lower chord splices of the Cheat River Bridge, do these bolts take any primary stress. The transfer of load from one element to another was accomplished by direct bearing or by notching. Wedges were used quite freely to take care of shrinkage conditions and wooden pins were utilized in the bracing systems.

Peter T. Landsem,[66] Assoc. M. Am. Soc. C. E. (by letter).—The authors deserve much credit for their thorough investigation of the development of wood bridges in the United States. However, the paper gives the impression of being an epitaph on wood bridges, which is not deserved. Certain developments in the more efficient use of lumber both in Europe and in the United States indicate that the world is probably on the threshold of a renaissance in wooden bridge construction.

In general, information on the period from the collapse of the Roman Empire up to 1500 A. D., seems to be lacking in the authors' treatise. Improvements were small, which is astonishing in view of the highly developed church architecture, both Romanesque and Gothic. Most bridges were simple trestles. An outstanding example was the one built over the Rhine, at Basel, Switzerland, in 1225, and first replaced in 1903.

It might be of interest to state that Palladio did not consider himself the inventor of the truss. He refers to a truss with parallel chords, verticals, and single compression diagonals found in use in Germany by Picheroni de Mirandola.

The authors state that "covered or roofed-in bridges were distinctly not a feature in European practice," yet there is evidence that covered bridges were known in Germany and Switzerland long before the famous Ulrich Grubenmann and Josef Ritter became active in bridge construction early in the Eighteenth Century. One of the oldest of such bridges from the Middle Ages, still in existence, is the picturesque "Kapellbrücke" over the Reuss, in Luzerne, Switzerland, built in 1333. Similar bridges were built in the Sixteenth Century with spans up to 22 m. In 1658, a covered bridge with a span of 32 m was built over the Rhine, at Stein-Säckingen, Germany. Dr.-Ing. K. Schaechterle,[67] has called attention to a number of covered bridges, and advice concerning the construction of covered bridges has been published by L. Chr. Sturm;[68] J. Leupold;[69] Johan Wilhelm;[70]; and Johann Vogel.[71]

[66] Asst. Engr., National Committee on Wood Utilization, U. S. Dept. of Commerce, Washington, D. C.

[67] "Holzbrücken," 1927.

[68] "Architectura civili-militaris," 1719.

[69] "Theatrum pontificiale," 1726.

[70] "Architectura civilis," 1682.

[71] "Die Moderne Baukunst," 1708.

Developments in Europe in the Nineteenth Century.—The authors of this treatise do not trace the development of wooden bridges in Europe after the middle of the Eighteenth Century, and do not cite any examples of such bridges built in the latter part of the Nineteenth Century in the United States.

A brief summary of the development in Europe may be of interest. About 1800, Funk, in Germany, applied to bridges a wooden arch that had been used for roofs by the French architect, de l'Orme, in the latter part of the Sixteenth Century. The arches were made of boards on edge, cut to shape, and nailed and doweled together. This system was not satisfactory, however, because in shaping the boards the grain of the wood was cut too much and because water entered so easily into the vertical joints.

In 1809, Wiebeking, another German, used for bridges the system generally attributed to the French Colonel Emy, although not in use by him until 1819. The boards were laid flat and nailed and bolted together. This type of arch, however, was too flexible even when braced. Load tests on both de l'Orme's and Emy's systems were carried out by the French military engineer, P. Ardant.[72]

Although the railroads increased the demand for bridges early in the Nineteenth Century, no significant improvements originated in Europe. It was left to the United States to develop the combined arch truss into the framed trusses of Long, Howe, and Pratt. From the middle of the century the art of wooden bridge building went into a decline. This must be attributed mainly to lack of knowledge concerning the properties of wood and concerning reliable framing methods. Engineers of that day lost interest in wood construction. In fact, the only new idea that has come to the writer's attention are the gallow bridges in Trondhjem, Norway.

Modern Developments in Wood Construction.—New interest in wood construction was created about 1900 by carpenter-contractors who, having acquired some technical knowledge, put up structures that caught the inter-

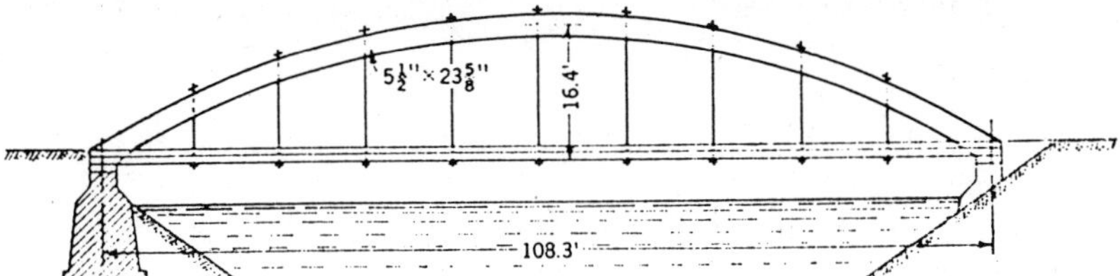

FIG. 39.—SECTION OF FOOTBRIDGE OVER THE WIESE RIVER AT BASEL, SWITZERLAND

est of engineers. The engineers have carried the developments further. Therefore, in Germany to-day, they speak about "Ingenieur-Holzbau," or engineering wood-construction.

[72] "Theoretischpraktische Abhandlung über Anordnung und Konstruktion der Sprengwerke von grosser Spannweite," von A. v. Kaven, 1847.

Stephan made Emy's arch stiffer by building it like a truss of curved laminated chords, doweled together, and braced by cross-diagonals. This system, however, was used only for roof trusses. The railway station in Copenhagen, Denmark, is a good example.

Otto Hetzer introduced his laminated glued-wood arch in 1907. It consists of boards laid flat, curved, and glued together, the glue being water-resistant and having at least the same strength as the wood. An early example is the foot-bridge over the Wiese, at Basel, Switzerland (see Fig. 39), which was built in 1910. The roadway is suspended from the two-hinged arches, 33 m (108.3 ft) in span. It was computed for a load of 350 kg per sq m (72 lb per sq ft) and cost 6 200 francs, which was only one-half the nearest bid in steel. Th. Gesteschi[73] and H. Lewe[74] mention other arch bridges.

FIG. 40.—VIEW OF CONSTRUCTION BRIDGE OVER THE SPREE, BERLIN, GERMANY: CLOSE-UP OF ONE SPAN

During the World War the simple truss was improved and the trussed frame was introduced with the use of new and more efficient joining devices. Modern European literature refers to many wood bridges. A covered bridge over the Neckar, at Thalhausen, Germany, was built in 1924. It has three spans, 62.3 ft long and two spans, 32.8 ft long. Others are: A suspension bridge at Sulitjelma, Norway, 266 ft long (maximum span, 177 ft); a pipe-line suspension bridge in Finland, with four spans of 167 ft each and a total length of 984 ft; a footbridge at Neumünster Railway Station, in Germany (maximum single spans, 131 ft, and total length, 541 ft); and a military

[73] "Der Holzbau."

[74] "Neuere Brückenbauten in Holz," *Bauingenieur*, 1921, Vol. 17.

road bridge in Norway (span, 85 ft). Well known foreign sources of information are: Th. Gesteschi;[73] C. Kersten;[75] and A. Laskus.[76]

Introduction of Modern Wood Construction Methods in the United States.—It has been recognized that the chief drawback in the use of wood to-day in bridge construction has been the inefficient joint. In bolted connections the safe loads determined by the various handbook methods differed widely. This situation, fortunately, has been relieved by an extensive series of tests[77] carried on for years at the Forest Products Laboratory, U. S. Department of Agriculture. The Engineering Staff of the National Committee on Wood Utilization, U. S. Department of Commerce, in co-operation with the Forest Products Laboratory, U. S. Department of Agriculture, has been investigating still better joining devices which, as tests show, may increase the load capacity from two to six times (and even more) than that of an ordinary bolted joint.

Fig. 40 shows a bridge in which modern connectors are used in the joints. These joining devices or connectors, comprising various types of rings, plates, and disks embedded in the sides of timbers (see Fig. 40), are used for transmitting the load from one member to the other. Fig. 41[78] is the view of a joint composed of modern connectors developed in Germany.

FIG. 41.—VIEW OF JOINT COMPOSED OF MODERN CONNECTORS

In concluding, the writer would like to call attention to the fact that the State of Oregon is building timber highway bridges, using the money thus saved for better road surfaces.

---

[75] "Freitragende Holzbauten."

[76] "Hölzerne Brücken."

[77] *Bulletin No. 332*, "The Bearing Strength of Wood Under Bolts," by George W. Trayer, 1932.

[78] "Modern Connectors for Timber Construction," by Nelson S. Perkins, Peter T. Landsem, and George W. Trayer, *Bulletin No. 24*, National Committee on Wood Utilization, U. S. Dept. of Commerce.

J. K. FINCH,[79] M AM. SOC. C. E. (by letter).—One or two books on American timber bridges have appeared in recent years, but this paper is the first to record and discuss some of the interesting technical details involved in early timber bridge construction. While the men who built these bridges were simply self-taught Yankee carpenters, their abilities were of the first order and from their labors the truss bridge, a typically American contribution to engineering progress, was evolved.

The writer has no details to add to the record of these master builders as given by Messrs. Fletcher and Snow. There are, however, a few notes that might be of interest on the place this timber-bridge era occupies in the great epic of engineering achievement.

The authors mentioned the lack of bridges in ancient Greece. The same observation would hold for Egypt and Mesopotamia. This was not due to any lack of ability on the part of the Greek architecton, or master builder. (The name, architect, is derived from this title but it is preferable to translate architecton, and the later Roman architectus, as master builder, for, until the introduction of a new technique with the advent of gun-powder in the early Renaissance, the present professions of civil and military engineering and architecture were all included in this single name). Topography was a most important element in shaping the course of engineering development. Thus, the Nile and the Tigris-Euphrates furnished the transportation routes for these two most ancient civilizations. Roads, and hence bridges, were not required. Again, Greece has been described as a huge, bony hand stretching out into the Eastern Mediterranean and Aegean Seas. The construction even of trails across her rock-ribbed peninsulas was difficult. A unified Greek nation was topographically impossible. Roads, as an important element of communication, were not to be thought of. The Greek City States turned to the sea. The first naval battle of the world was fought by the Greek and Persian fleets at Salamis, and the Greeks naturally became the first of ancient harbor builders.

The topography of the Roman peninsula, on the other hand, made it physically possible to consolidate this great area and its many tribes into a single nation from the Alps to the heel of Calabria. This opportunity was taken advantage of by the strongest tribe, the Latin race, and the construction of roads, and necessarily of bridges, became an inevitable, a basic element, to this consolidation. The Alps shut off France from Italy, but France also was consolidated by another great Roman road and bridge-building development, while communication between the seaports of Southern France and Northern Italy furnished the connecting link.

Centuries later, in the United States, economic pressure from the seaboard States, the struggle of each State to secure trade advantages from the expansion westward which followed the Revolution, combined with the topographic problems involved, made the American canal and railroad era, and its accompanying bridge development, inevitable. It was not only Yankee ingenuity, or the force and initiative of a pioneer people, that made the United States

[79] Renwick Prof. of Civ. Eng., Columbia Univ., New York, N. Y.

the greatest nation of bridge builders the world has ever known. Topography and economic pressure have forced the people of this country to be ingenious and progressive in this particular field.

It is also interesting to reflect on some other factors and forces that led American engineers to develop the truss bridge.

Palladio's famous timber truss of about 1540, mentioned by the authors, is well known. The first metal truss, however, had been built centuries earlier, under the Roman Engineer-Emperor Hadrian, to support the portico of the Pantheon at Rome. Undoubtedly, the truss form was used by the Greeks and, later, by the Romans in many roof constructions. Timber, however, was recognized as temporary, and the fire hazard was also considered. Hadrian's truss, for example, was of bronze and was clearly a pioneer attempt at fireproof building construction. (It was removed and melted for making cannon under Pope Urban VIII in 1625, but early drawings show its form and details.) Timber bridges, although undoubtedly used, were replaced by fireproof, durable stone arches whenever economic conditions warranted permanent construction. Hadrian's effort to build framing in metal was interesting, but of no economic importance because metal was too costly. In fact, the truss, an invention in framing, did not emerge, was not discovered or invented, until a civilization arose which was based almost entirely on timber construction. Colonial America was such a civilization.

David Stevenson, the British engineer, writing[80] in 1838 defended the use of timber and temporary constructions by American engineers, a practice which had been criticized by European workers. He pointed out that labor was scarce in the States, funds were not available for costly stone constructions, and the unknown future requirements as to loads and sizes of structures made temporary construction advisable. Construction in timber, therefore, represented not only economical design, but also good engineering and business foresight.

How much such pioneers in this field as Palmer, Burr, and Wernwag were indebted to earlier European bridge builders for their basic ideas, will probably never be definitely known. "Copying" may not have occurred at all. Indeed, books were scarce, and American Colonial carpenters would not be expected to have or see them. This point, however, is unimportant. These men and their associates, as well as earlier workers, apparently had no very clear conceptions of truss action, or at least did not have full confidence in such framing. Their bridges were "statically indeterminate" combinations of arches with "stiffening" trusses. Nevertheless, the germ of the truss principle existed in their work, and ultimately the truss emerged from their composite structures as a form capable in itself of carrying the full load. Town, Long, and, finally, Howe (1840) saw this element in the work of Palmer, Burr, and Wernwag, and the truss was thus brought forward as a distinct and separate structural form.

During this period of truss evolution, British engineers stuck stubbornly to their metal arches, girders, and "tubular" bridges. American engineers,

[80] "Sketch of the Civil Engineering of North America," pub. by John Weale, Lond., 1838.

following Haupt and Whipple, began to analyze the stresses in trusses. There was an era of patented trusses in which Pratt, Warren, Bollman, and others contributed to the evolution of economic truss forms. The timber and wrought iron was translated into the cast and wrought-iron construction of the Sixties, and then into the all-steel truss of the Seventies and Eighties. Finally, the engineers of the present day have contributed the great alloy steel spans of the modern highway era.

The work of the early American timber bridge builders, therefore, represents only one step in the history of a great development. Their contribution to engineering progress, however, entitles these pioneer American bridge engineers to a prominent place in the annals of a great profession.

A. A. Eremin,[81] Assoc. M. Am. Soc. C. E. (by letter).—A number of ancient wooden bridges have been described (in Russian) by Mr. L. Nikolai,[82] a reference to which should be made in connection with this paper. For example, he shows the Trajan Bridge over the Danube River at the Iron Gates built by Apollodorus about 109 A. D. This is the bridge to which reference was made by the authors. The clear span was 118 ft, and the stone piers were carried 20 ft below water level. The Trajan Bridge is known as the first wooden arch bridge.

A freedom of the creative imagination at the beginning of the Nineteenth Century is well illustrated by Mr. Burr's bridge,[31] at Schenectady, N. Y., built in 1808. The curved chords were formed of eight 4 by 14-in. planks, spiked, and bolted together. After twenty years of service the bridge was reinforced with additional intermediate piers built in the middle of each span, and, in 1873, the structure was replaced by a metal bridge.

Among the notable wooden bridges built in Russia, Mr. Nikolai has described the "rainbow" arch bridge with suspended deck and a clear opening of 257 ft, spanning the Wepr River at Ivan Gorod Fortress.

In crossing navigable rivers the Russian builders used a removable bridge with draw-spans that were easily opened. The bridge was removed in winter when the river was covered with ice, and in spring, to permit a heavy mass of floating ice to pass at high water.

A primitive type of removable bridge is the floating type of structure. The deck of a floating bridge built in Riga (Fig. 42), was formed of planks

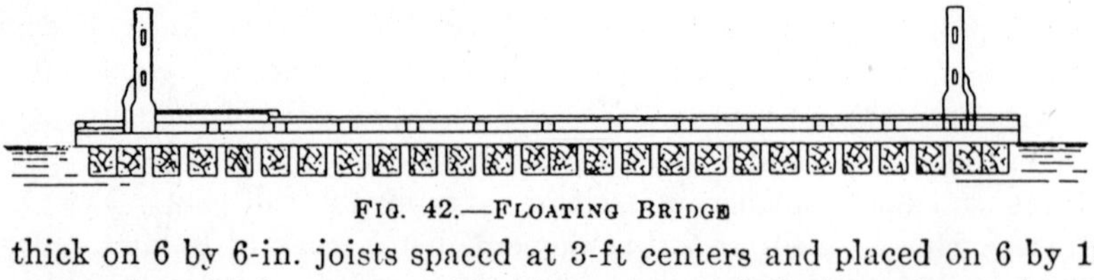

Fig. 42.—Floating Bridge

3 in. thick on 6 by 6-in. joists spaced at 3-ft centers and placed on 6 by 12-in. beams spaced at 3½-ft centers. The beams were bolted to the 13 by 13-in. floating longitudinal beams. The average life of this type of floating bridge was about ten years. Later, the floating bridges were replaced with pontoon bridges, the spans being carried on wooden or metal barges.

[81] Asst. Designing Engr., Bridge Dept., State Highway Comm., Sacramento, Calif.
[82] "Bridges," by L. Nikolai, Russian Edition, 1901.

AKSEL ANDERSEN,[83] M. AM. SOC. C. E. (by letter).—The following data on wooden railroad bridges built in Norway between 1860 and 1875 is of interest in connection with this instructive paper on the development of wooden bridges. During that period the Norwegian Government constructed several important railway lines, including the northern half of the railroad between Oslo and Trondhjem, and several lines in the southeastern part of the country.

A great variety of conditions were encountered, involving the bridging of numerous rivers and deep valleys in mountainous country, with poor transportation facilities. Skilled carpenters and excellent pine timber were available, however, at low cost, and the selection of wood as the principal building material was natural. Stone or iron bridges would have been much more expensive and would have required a longer construction period.

FIG. 43.—VIEW OF GULFOSS BRIDGE

In the lowlands the pile trestle was a common type, with spans as great as 40 ft. Compound keyed stringers with queen or king-braces were used for the longer spans. Practically all structures of this type were built of round, or half-round timber, and the joints were held together with iron screw-bolts.

For river crossings and high viaducts, with spans up to 100 ft, Pratt trusses and Howe trusses, with one and, in a few cases, with two systems of web members were used. Fig. 43 shows a bridge (Gulfoss Bridge), with four continuous-truss spans of 100 ft and several shorter spans. The continuous trusses are of the original Howe type with two web systems, very similar to

[83] With Ash-Howard-Needles & Tammen, New York, N. Y.

the trusses of the bridge over the Connecticut River, at Springfield, Mass., shown in Fig. 24. The only apparent difference is the absence of braces against the piers below the lower chord. The same system of continuous 100-ft spans, with two systems of web members, was used in two other railroad bridges, and one of them (the bridge over the Gula River, at Stören), is still in use (1933) after nearly seventy years of service. (In this connection note the statement by Messrs. Fletcher and Snow under "The Howe Truss," that "it is not known to the writers that there has been another bridge built of the Howe type with webs of two systems"). The Gulfoss Bridge and several other through railroad bridges were protected by roofs, extending about 6 ft outside the trusses, with no side coverings.

Deck spans with simple Howe trusses, or Pratt trusses, supported by timber piers on masonry bases, were used in a number of high viaducts. The Dröja Viaduct, shown in Fig. 44, had one 100-ft, and seven 45-ft, truss spans of the Pratt type. The upper chords and the post were made of wood, but the diagonals and the lower chords were of iron. The diagonals of the 100-ft span consisted of two rods varying from $2\frac{1}{8}$ in. in diameter at the ends to 1 in. at the center. The diameter of the threaded ends of the rods was $\frac{1}{4}$ in. larger than that of the rod; the threaded sections (6 in. long) were attached to the rods by welding. The welds were, obviously, points of weakness.

The wind system consisted of timber crosses in the plane of the upper chords with iron tie-rods at the panel points, and timber cross-frames at each panel point with iron tie-rods at the lower chords. The writer recalls an incident, in about 1916 or 1917, that proved the efficiency of this type of wind stiffening. One of the iron diagonals in a Pratt truss of a high timber viaduct failed, and several trains had probably passed over the viaduct before the break was detected. The only reason why the structure did not collapse was that the load was carried by the other truss and the wind system. A few typical details are shown in Fig. 44.

The Dröja Viaduct and other deck bridges and viaducts of the same type, were protected by a wooden roofing, extending about 3 ft beyond the outside faces of the upper chords. The trough-shaped part between the rails was made of longitudinal boards, covered with iron plates in order to reduce the fire hazard. The roofing outside the rails consisted of two layers of 1-in. by 6-in., transverse boards sloping slightly outward. The boards in each layer were placed with a spacing of 2 in., thus overlapping the boards of the other layer by 2 in. The contact surfaces between the two layers were provided with grooves for drainage (see Fig. 44). A similar system was used for wooden floors of highway bridges. The wearing surface consisted of longitudinal planks, and the loads were carried to the stringers by two layers of overlapping and grooved planks.

Simple Howe trusses, with iron rods as verticals, were used in a number of railroad bridges and viaducts. In some of them, the angle blocks were made of wood (usually oak), but more frequently they were of cast iron.

Practically all wooden bridges designed and built by the Norwegian Government Railways in the aforementioned period have been replaced, in recent years, by stone or steel bridges, or have been eliminated by re-location

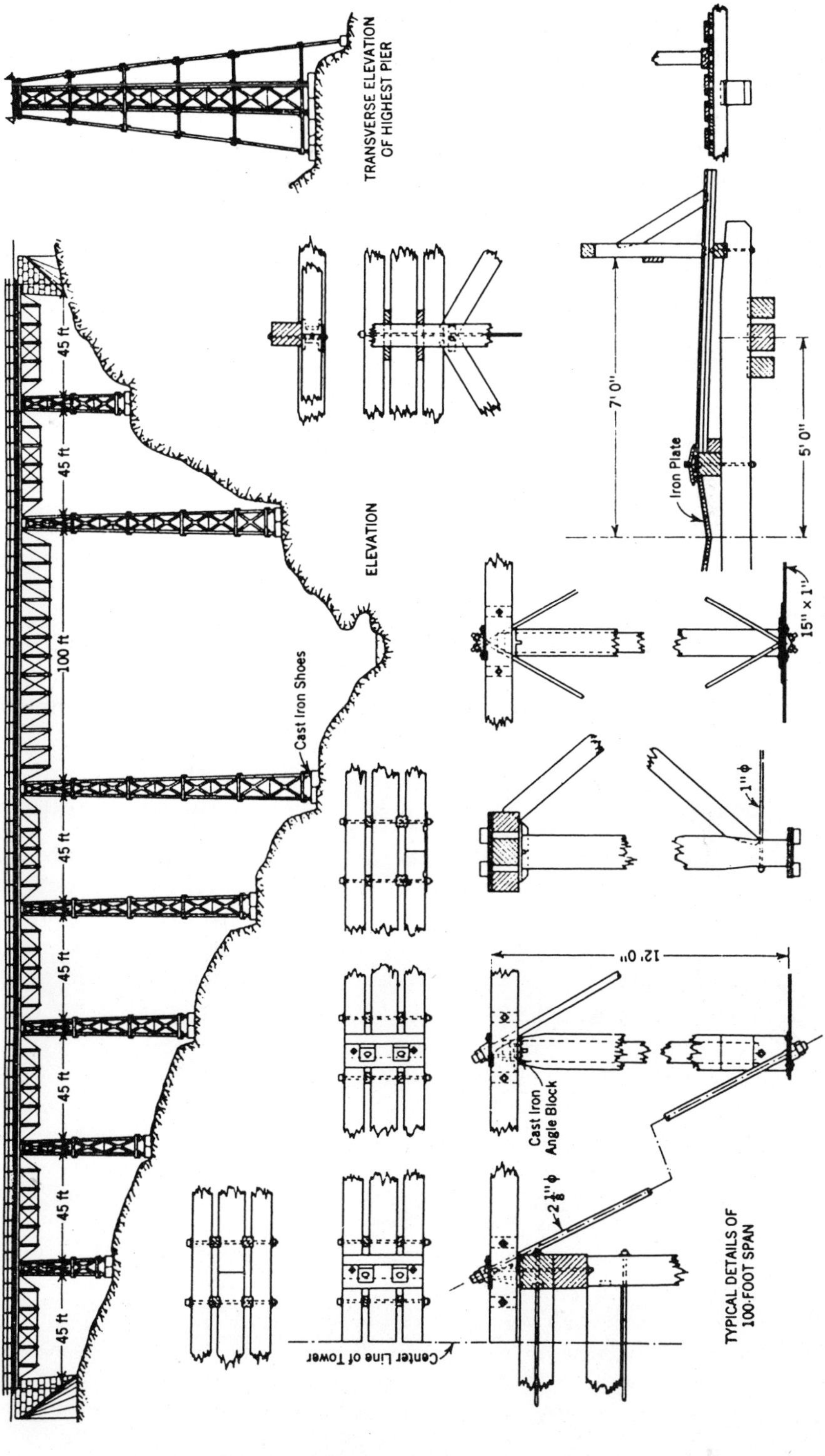

FIG. 44.—TYPICAL DETAILS OF TRUSS SPANS, DRÖJA VIADUCT

of the lines. Record drawings, based on field measurements, and a number of photographs of the more important bridges have been collected by the Railway Museum at Hamar.

The writer wishes to acknowledge his indebtedness to Hans Tönnessen, Chief Engineer of Bridges of the Norwegian Government Railroads, who furnished drawings, photographs, and other data for this discussion.

William G. Atwood,[54] M. Am. Soc. C. E. (by letter.)—This paper has great historical interest and contains many lessons for engineers in active practice to-day. The interest created by the development of new or improved materials and the very intelligent advertising and salesmanship of their producers have caused the virtues of some of the older materials to be disregarded.

It is the duty of the engineer to build the cheapest structure that he can, which will be fully adequate for its purpose. If he studies this paper carefully he is likely to question whether many times a full timber or a composite structure, when taking into account interest on investment, maintenance, probable obsolescence, and salvage value, is not really a better and more economical structure than one built entirely with the newer materials.

Most of the structures described by the authors were built of timber because it was the only material physically available. This is still true to-day in a few isolated cases. The writer had a part in the construction of the Alaska Central Railway (later, the U. S. Government Railway), on which, in a distance of 4 miles, there was a total of more than 1 mile of timber bridges, part trestle and part Howe truss. This work was done in 1906 and 1907 when no transportation except pack and wagon train was available and no other material could have been used. Examples of this kind are rare in the present day.

Under many conditions timber trestles or spans will be found more economical than concrete or steel if all factors of cost are taken into consideration, especially that of obsolescence or the possibility of future reconstruction or widening. As an example of this there may be mentioned a number of short-span, through, concrete girder highway bridges on one of the most important north and south through highways. These bridges were built a number of years ago and roadway widening became necessary. The old bridges had to be removed and were replaced by others of the same type. If, originally, they had been treated-timber pile bridges, their first cost would have been less, they could have been widened at a small fraction of the cost of the work actually done, and the treated-timber bridges would have lasted as long as bridges were required in these locations.

This paper not only shows the great ability of the preceding generations of engineers in using the structural materials available to produce bridges that were not only adequate for the loadings of their time, but capable of safely carrying much greater loads, but it also shows the inherent value of the material itself. Timber bridges carrying loads several times those for which they were designed for periods of 50 to 100 years cannot by any stretch of the imagination be called temporary structures.

---

[54] Cons. Engr., New York, N. Y.

When giving credit to their professional predecessors engineers should remember that they did not have the information that is available at the present day even regarding such an old structural material as timber; nor was treated timber available in any quantity until the last fifty years. As a result of the work of the U. S. Forest Products Laboratory, the Department of Commerce, and the various associations of timber producers and timber treaters, together with the engineering societies, there is available for the present-day designer information that permits him to select his timber and treatment with a certainty as to strength and durability that is not exceeded in the case of other structural materials. Engineers of this generation, therefore, may build stronger and more durable structures with less timber than their predecessors. The paper under discussion shows what the past generation could do.

It is a fact which will probably be questioned by few who are conversant with present practices, that frequently advantage is not taken of the information now available. The Highway Department of one important State is now using timber grading rules and specifications written thirty years ago, and an important city is using creosoting specifications nearly as old. This is a waste of money.

Men like Timothy Palmer whose bridges are described in the paper, were in advance of their time. This generation of engineers may well ask themselves whether in their frequent disregard of the oldest and one of the best of the available structural materials, they are worthily "carrying the torch of progress."

H. Fugl Meyer,[85] Assoc. M. Am. Soc. C. E. (by letter).—A brief statement as to the contribution of the Chinese to the art of bridge building is of importance in establishing the origins discussed in this interesting paper. The Chinese character, "chiao," denoting a bridge, contains in its components the idea of a lofty wooden construction. The character was formed about 1003 B. C., thus lending authority to legendary history, which records the information that the Chinese had undertaken the construction of fairly large wooden bridges at a very early date. From the Confucian annals and from the history of the Near East it is known that in a number of Asiatic countries wooden pontoon bridges of considerable length were in regular use before 600 B. C.

At that early date there were a number of communities with a fairly high civilization in the part of Central Asia that now (1933) forms the western border of China. Because the turbulent and wide mountain streams in this territory made all inter-communication difficult, Nature made the inhabitants bridge builders long before written records of their kingdoms begin.

A number of interesting wooden cantilever and suspension bridges are still to be found in these regions. Whenever the width of the stream is less than, say, 130 ft, a cantilever appears to be preferred. Suspension bridges of plaited bamboo ropes are used for spanning streams as wide as 250 ft.

---

[85] Mgr., Free Port of Copenhagen, Copenhagen, Denmark.

Multiple-span bridges of both types of construction are still extant, the longest being a suspension bridge of 700 ft.

Chinese bridges are generally the principal gathering place for the inhabitants of the neighboring town, and, therefore, they are often covered with a roof or are fitted with supports for a light roof of bamboo matting as a protection against sun and rain. Fig. 45 represents a bridge in the small

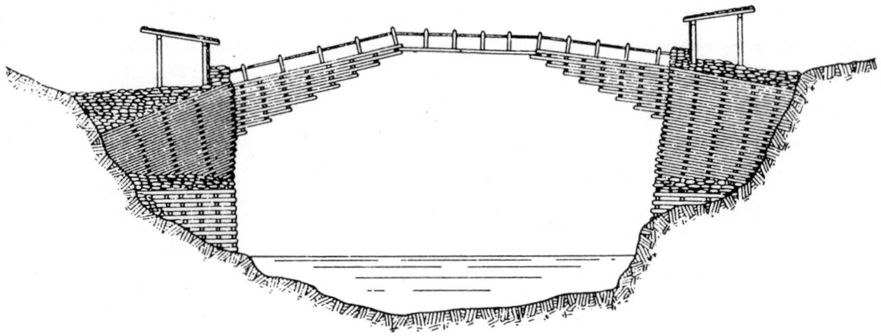

FIG. 45.—CANTILEVER BRIDGE IN THE KINGDOM OF MULI.

Tibetan-Chinese Kingdom of Muli. It will be noted that two heavy stone-filled wooden caissons lend stability to the cantilevers. The central part of the span can be removed when the King of Muli finds it necessary to block this entrance to his realm.

W. E. HOWLAND,[86] ASSOC. M. AM. SOC. C. E. (by letter).—In an "Encyclopædia or Dictionary of Arts, Sciences and Miscellaneous Literature," printed in Philadelphia, Pa., for Thomas Dobson, in 1798, and in the "Supplement" published in 1803, several articles appeared relating to wooden bridges, namely, "Arches," "Centering," "Carpentry," "Roofs," and "Strength of Materials," all related one to another with numerous cross-references. This twenty-one volume compilation[87] offers an opportunity for the student of the history of the development of technology to determine what was known about a particular science at a given period, and, in this case, it shows what knowledge might have been carried over from Europe to America. Such a medium as this could have brought to the early American bridge builders the knowledge gained by their European colleagues, not only in building wooden bridges, but also in the science of the equilibrium polygon and in countless practical details of wood construction for carrying heavy loads.

Although, as the authors have pointed out, the European wooden bridges were not especially noteworthy at this time, certain other Continental wooden construction appears to have been developed to a high degree of perfection before the era of American wooden bridge construction, for example, the wooden centering of Perronet and the wooden roof trusses of Sir Christopher

[86] Asst. Prof. of Civ. Eng., Purdue Univ., Lafayette, Ind.

[87] Reprinted from "Encyclopaedia Britannica," or a Dictionary of Arts, Sciences and Miscellaneous Literature," Third Edition, printed for A. Bell and C. Macfarquhar, Edinburgh, Scotland, 1797.

Wren. The writers of this encyclopædia assembled this knowledge and made it available in a convincing and understandable form for the potential bridge builder. It seems inconceivable that a reader of these articles would not have profited immensely both in obtaining a clearer mental picture of the principles involved and the manner in which the forces distribute themselves in a complicated frame, but also in adding to his knowledge of practical details of construction.

The article on "Strength of Materials" is a mine of information to the historian of science. Its author considers tension, compression, and shear and the resistances of various materials (particularly wood) thereto. He considers Euler's column theory, and he derives a practical rule for estimating the strength of a beam in bending which is in close agreement with modern ideas; that is, he assumes that the strength varies directly with the breadth of the beam and with the depth squared, and that a beam loaded in the center requires a strength that is proportional to the span and to the load. All the theory presented is discussed in the light of practical knowledge and is translated into rules-of-thumb for the use of the artisan.

The article on "Roofs" contains a discussion of improper practice in joining wooden members, and it ends with the maxim of Perronet, namely, "to make all the shoulders of abutting pieces in the form of an arc of a circle having the opposite end of the piece for its center."[88] Then, to connect this subject of roofs with other phases of wood construction, the author adds this sentence,[89] "much of what has been said on this subject may be applied to the construction of wooden bridges and the centers for turning the arches of stone bridges."

The work of Perronet is discussed in the article on "Centering,"[90] with critical insight and yet with admiration. This section treats of the history of the development of the center to produce the stiffness needed for maintaining the shape of the stone arch in various stages of its construction. The author commends the center used by Mr. Robert Mylne (1734-1811) for the Blackfriars Bridge, at London, England. This bears a striking resemblance to a Bollman truss, with the compression members and tension members reversed. The construction consists of a number of pairs of struts, each one beginning at an abutment and terminating at a joint on the outside of the centering, the entire framework being joined together to form a structure of great rigidity.

In the article on "Carpentry,"[91] the author again refers to the work of Perronet. He states, for example, that:

> "The strongest of all methods of piecing a tie-beam would be to set the parts end to end and grasp them between other pieces on either side * * *. Mr. Perronet used it for the tie-beams or stretchers by which he connected the opposite feet of a center which was yielding to its load and had pushed aside one of the piers above 4 in."

[88] "Encyclopædia or Dictionary of Arts, Sciences, and Miscellaneous Literature," 1798, Vol. 16. p. 476.

[89] *Loc. cit.*, p. 480.

[90] *Loc. cit.*, Supplement, 1803, Vol. 3, p. 234.

[91] *Loc. cit.*, p. 212.

The author then describes the use of scarfing employed by Perronet for eliminating bending in the bolts that connect wooden members. Two full pages of these and other joints are included, and two pages of illustrations of roof trusses, one with a clear span of 80 ft, and a sketch of the walking beam truss used by Watt and Boulton on their steam engine. It has tension members for the top chord.

The author on "Centering" discusses disapprovingly the wooden arch built by Grubenhamm, at Wittingten, Switzerland.[92] He then refers with admiration to a "bridge over the Portsmouth river in North America, * * * more than 250 ft in length * * * of several parallel arches of beams." The inventor said to be a Mr. Blodget, is quoted as saying that he found the strength of this bridge so great that he could with perfect confidence make one of four times the span. Could this be the Piscataqua Bridge of Timothy Palmer built in 1794 and referred to by Messrs. Fletcher and Snow?

Finally, in recommending the best form for a wooden bridge, the contributor of this article returns to the ideas of the roof truss and the arch centers, concluding with these significant words:[93]

"The reader must perceive that we have now terminated in the construction of the Norman roof. We indeed think it the best general form, when some moderate declivity is not an insuperable objection. When this is the case, we recommend the general plan of the centering of the bridge of Orleans. We would make the bridge (we speak of a great bridge) consist of four trusses; two to serve as the outsides of the bridge, and two inner trusses, separating the carriage-way from the foot-paths. The road should follow the course of the lower polygon, and the main truss should form the rails. It might look strange; but we are here speaking of strength; and evident, but not unwieldly, strength, when once it becomes familiar, is the surest source of beauty in all works of this kind."

HARRY K. ELLIS,[94] ESQ. (by letter).—The authors state that "as late as 1930, there were between 450 and 500 covered bridges in use." This must be an under-estimate, because in Chester County, Pennsylvania, alone there are forty-four covered wooden bridges at the present time. Nine were torn down in 1932, so that in this County there must have been at least fifty-three covered bridges in 1930. Many of the other counties in Eastern Pennsylvania still have large numbers of such bridges.

Among the bridges still existing in Chester County is one of 100 ft span, built in 1807 and still carrying traffic, and one of 60 ft span, built in 1819. The latter bridge is shored up at the present time. The writer believes that the one built in 1807 must hold the record for long life of Burr truss bridges, since Burr's first bridge was built only three years earlier.

Most of the covered bridges in Chester County are of the Burr type with the wooden arches extending to a height about 2 ft below the eaves. White pine was used in almost every case. Some of the bridges are of the queen-post type, without diagonals in the middle panel, and with the floor-beams

[92] "Encyclopædia or Dictionary of Arts, Sciences, and Miscellaneous Literature: Supplement," 1803, Vol. 3, p. 248.

[93] *Loc. cit.*, p. 249.

[94] County Engr., Chester County, West Chester, Pa.

hung upon a single bolt which passed through the center of the floor-beam upward through the lower chord and about 1 ft to 1.5 ft higher through the truss post. A notch cut in the post permitted the insertion of a nut into which the bolt was screwed. Usually, this notch was on the outside of the truss and covered with the siding boards so that inspection was difficult.

It is true that the floors were the weakest part of these bridges, but the writer has twice seen a perfect compression failure of the arch rings. In both cases the wood broomed out on all four sides of the ring at the point of failure. In both cases, the arch was subject to a greatly increased load due to the bridge as a whole being much out of line.

NELSON J. BELL,[95] M. AM. SOC. C. E. (by letter).—This paper is most instructive, especially since technical literature in the past half century seems to have almost wholly neglected the subject. The general opinion among engineers has come to be that the old wooden highway bridges are fit objects of sentimental interest, but of very little practical worth in this age of steel and concrete. The paper by Messrs. Fletcher and Snow should bring home the fact that with only nominal care these old bridges have served their purpose with almost 100% efficiency over periods of 50 to 100 years, and, therefore, may be worthy of close technical study before it is decided whether one should be further preserved for practical use or should be condemned and scrapped. The mortality of these old bridges has been greatly increased in recent years because of the lack of knowledge concerning them and the consequent offhand opinion that they are unsafe under present-day traffic.

The preponderance of evidence brought out in this paper is to the effect that the bridges built by Wernwag, Burr, and most of their imitators, had an excess of strength everywhere except in their floor systems and lateral bracing, but no one has as yet described a method by which the distribution of stresses in the trusses may be analyzed. An exact analysis is probably impossible, and, in any case, would involve so many unknown factors that it would be highly impractical. In some cases it is possible, however, to arrive at a reasonable analysis of the primary stresses in the members by means of assumptions and simple graphic statics.

Several years ago, the writer was employed to investigate a wooden bridge spanning the Kentucky River at Camp Nelson, about twenty miles south of Lexington, Ky. This bridge was built in 1838 by Lewis Wernwag. It has a single span of 240 ft and is composed of three trusses which are identical in all respects, except that the outer ones are curved horizontally so as to form flat arches, equal and opposite to each other. The outer trusses are spaced about 25 ft, center to center, at the middle of the span, and about 30 ft, center to center, at the ends. The center truss is straight and bisects the bridge, forming two one-way roadways. The general design is so similar to that of the bridge over Cheat River, Preston County, West Virginia (Fig. 35), that both must have been designed by the same man. The Camp Nelson Bridge is much heavier, the arches being composed of six pieces, in two rings of three

[95] Structural Engr., Dayton, Ohio.

pieces each. There is also a double web system, the main diagonals being composed of two pieces each, passing outside the arch rings, and the single-piece counters passing between the two arch rings. The verticals have three pieces each, the center one passing between the arch rings and the others passing outside. Thus, the arches are clamped against lateral deflection by the verticals and diagonals. The verticals are dapped over the arches and bolted to them, but the diagonals have no connection to the arches, except by friction. An illustrated description of this bridge, with a tabulation of maximum stresses and an outline of the method of analysis was published[96] in 1928 by the writer and J. K. Grannis, M. Am. Soc. C. E.

It seems to be characteristic of Wernwag's bridges that they have heavy arches with comparatively light stiffening trusses, and that their ends are anchored to the abutments by tension rods. The Camp Nelson Bridge is composed essentially of three arches, each braced laterally and vertically by a stiffening truss, and the whole tied together with lateral trusses at top and bottom. The points of application of loading on each arch are spaced uniformly along the perimeter of the arch by means of the inclined posts, set normal to the curve of the arch. Dead and uniform live loading, therefore, causes compression in the arch, coincident with its center line. The truss is not stressed except by eccentric loads, under which condition it distributes the loads uniformly to the arch, thus limiting all stress in the arch to compression. The stress distribution can be reasonably compared to that in a suspension bridge, the only difference being that in this case there is an arch in compression instead of a cable in tension. With any eccentric load on the truss, it tends to rotate on the arch. This must be counteracted either by the dead weight of the truss, or by anchoring the ends to the abutments, and it is interesting to note that the trusses of the Camp Nelson Bridge were actually anchored by wrought-iron rods extending from the tops of the end posts down into the masonry.

The trusses by Burr and some others do not lend themselves to stress analysis as readily as those by Wernwag, as they seem to have been designed on the principle of a simple truss aided and strengthened by arches, so that there is no definite place where truss action stops and arch action begins. It would certainly be on the safe side to make the analysis on the basis of a stiffened arch where the arches have adequate lateral and vertical support to prevent distortion, but as the dead loads of these bridges greatly exceed the live loads, it is more practical to assign the dead load to the arches and the live load to the trusses and analyze them separately. In any case, the judgment of the investigator enters largely into the problem, and must supplement the mathematical work.

The main point which the writer wishes to make is that the old wooden bridges should not be condemned and scrapped just because they were not designed with the aid of present-day engineering knowledge and standards. Many of them can be made safe and serviceable for modern traffic by very simple repairs and reinforcement of their floors and lateral bracing systems, at a small fraction of the cost of removal and replacement.

[96] *Engineering News-Record*, Vol. 100, No. 6, February 9, 1928, p. 234.

ROBERT FLETCHER[97] AND J. P. SNOW,[98] MEMBERS, AM. SOC. C. E. (by letter).—A gratifying interest is revealed by the number of discussions submitted in connection with this paper. These discussions have brought to light historical matter of great value, and have increased the knowledge of timber bridges still doing good service. They have widened the view of the ancient history and of the later European period. That timber bridges have yet a "reason to be" is amply proved; and their use for some decades seems to be possible. These fruitful results of a suggestion made by other members of the Society constitute an important chapter of the history of the profession, which might have been lost beyond recovery.

Concerning Cæsar's military bridge—to meet the objection that the notched blocks, as the only connectors at the end of the bents, might fail by crushing at the sharp edges, or by splitting between the notches—a model of one span on two bents was made on a scale of one-twelfth that of the prototype. An estimated load on the span to include the weight of the large, rough-hewn green timber and a column of troops, four abreast, carrying arms and accoutrements, was computed to exceed 14 000 lb. By scale reduction the equivalent load on one bent of the model should be in excess of 100 lb (1 to 144). The model sustained two heavy men aggregating nearly 400 lb in weight, on two bents; that is, nearly 200 lb on one bent. After separation, the parts gave no indication of over-strain.

Figs. 46 and 47 are two notable views of Timothy Palmer's Bridge at Easton, Pa. (see Fig. 11), taken after the bridge had been in continuous service for 89 years. These photographs were taken in 1894 by Eugene Mowlds, M. Am. Soc. C. E., just one year before the removal of the bridge to make way for a structure capable of sustaining traffic by trolley cars. These views of the exterior are valuable as showing the care which the master builder exercised to produce a distinctive architectural effect, in contradistinction to the barren, cheap-as-possible housing since adopted in covering wooden bridges. Fig. 46 shows the appropriate covering with pleasing lines, harmonizing with the surroundings and giving a sturdy aspect to the structure as a whole. Fig. 47, a view of the Easton end, shows enough adornment to relieve homeliness; and, above the portal, is the name and date to serve as the signature attesting the authorship.

Under the heading "The Howe Truss," the writers mention as unusual types, two Howe truss bridges with an arch on both sides of each truss, and state that they had no knowledge of any other Howe truss bridges so constructed. In correspondence with the writers, F. J. Nevins, Valuation Engineer, of the Chicago, Rock Island, and Pacific Railroad Company, and Frank F. Fowle, of Frank F. Fowle and Company, Consulting Engineers, of Chicago, Ill., have stated that the first bridge over the Mississippi River was built for the Chicago, Rock Island and Pacific Railroad Company, between Rock Island, Ill., and Davenport, Iowa, during the years 1853 to 1856. It consisted of five fixed spans of Howe trusses, each about 253 ft, with arches on both sides of each

[97] Director Emeritus, Thayer School of Civ. Eng., Dartmouth Coll., Hanover, N. H.
[98] Cons. Engr., Boston, Mass. Mr. Snow died on September 4, 1933.

FIG. 46.—BRIDGE OVER DELAWARE RIVER, AT EASTON, PA.

FIG. 47.—ENTRANCE TO BRIDGE OVER THE DELAWARE RIVER, AT EASTON, PA.

truss, and a draw-span. They also state that a second bridge was built a few years later on the same masonry, of Howe trusses having the unusual feature of curved upper chords, and no auxiliary arches. These are historic bridges because of the strenuous and prolonged opposition of the steamboat owners up and down the river to their construction, and because the unbroken railroad connection thus made across the river gave a great impetus to the settlement of the States beyond.

Questions listed in a letter by J. R. Worcester, M. Am. Soc. C. E., with appropriate answers by the writers, are as follows:

Question 1.—Have not arches been found beneficial to abutments, as well as to trusses, in furnishing lateral pressure where most needed?

Answer.—An arch, of course, tends to assist an abutment in resisting the pressure from the filling at its back, but it should not be depended upon to save a weak abutment. An instance is recorded in which a bulging abutment led to collapse of the arch, breaking it at the crown, and buckling the truss. On the other hand, a case is cited of a **T**-abutment with narrow stem, in which buttress masonry behind the heads was needed when arches were added to the bridge.

Question 2.—What is the reason for the different thicknesses of planks shown in Fig. 17?

Answer.—It is wholly to facilitate construction. The customary way to start an arch is to set up light supports on the falsework to define the curve of the intrados and tack $\frac{7}{8}$-in. boards, which bend easily, to them; after which, 2-in. plank (perhaps not so many as shown), are secured to get a solid support on which the 3-in. or 4-in. planks can be clamped and spiked.

Question 3.—Does the working between parts in combined arches and trusses show itself in loosening of joints, in course of time?

Answer.—In lattice bridges the connection between arch and truss is invariably made by vertical rods through the arch and floor-beams or by needle-beams below the lower chords. Any slackness in these rods is corrected by the screw-threads at their ends. This type of connection is advisable with other kinds of truss, but in the Burr trusses of early days the favorite method was to frame the posts to make seats on the arch. The flexibility of timber is such that no serious working at these joints has been observed by the writers. The creaking observed when a train crosses one of these bridges is normal to wooden bridges. It may be caused by warped sticks coming to a full bearing when a load passes over the bridge. It is probably more pronounced in spruce bridges than in those built of white pine, on account of the greater tendency of spruce to warp.

Question 4.—Can you give any figures as to sag of a wooden truss due to shrinkage, and is this less in a Town lattice than in a Howe truss?

Answer.—Sag in wooden bridges is probably not so much due to shrinkage as to compression of surfaces in contact at joints. Lattice trusses without arches naturally sag more than Howe trusses on account of the great number of joints composed of wooden pins in bored holes. This involves pressure on the rough side-grain of pins as well as a tendency to deflection of the pins. To counteract this tendency it is good practice to build 50% more camber into lattice trusses than into those of the Howe type. Howe trusses, if sagged, can be adjusted by setting up their vertical rods. In the lattice type the sag is cumulative. The unique bridge at North Blenheim, N. Y., shown in Fig. 20, has held its camber from 1855 to the present, although it has no vertical truss rods. This is due, without doubt, to the care used in its framing and the seasoning of the timber before framing as noted by Mr. Morse in commenting on the Ninth Street Bridge, in Pittsburgh, Pa.

Question 5.—What units have you found safe for stresses in bolts and treenails in shear? For direct tension and horizontal shear in spruce and yellow pine? Also, in fiber stress in stringers?

Answer.—As is shown in Mr. Guppy's discussion the size of members depends in some cases on connections rather than on longitudinal stress. Unit stresses that have been found, by experience, to result in good practice are listed in Table 1.

TABLE 1.—ALLOWABLE STRESSES IN TIMBER

Description	Yellow pine	Spruce	Oak pins	Iron
Tension	1 000	800	.....	......
Compression	800	650	.....	......
Bending modified by factor containing depth ÷ span	1 200	1 200	.....	......
Bearing end-grain	1 000	800	.....	......
Bearing side-grain, whole width of timber	350	300	.....	......
Bearing side-grain, small area like washers	400	350	.....	......
Shearing, longitudinal	80	80	.....	......
Bearing	.....	.....	400	......
Shearing	.....	.....	600	......
Bending	.....	.....	1 800	......
Floor-beam hangers; tension, net section	.....	.....	.....	7 500
Lower chord coupling rods; tension, net section	.....	.....	.....	10 000
Lower chord gib-plates; bending	.....	.....	.....	12 500
Arch suspender rods; tension	.....	.....	.....	10 000
Lateral rods, loaded chords; tension	.....	.....	.....	15 000
Lateral rods, unloaded chords; tension	.....	.....	.....	20 000

Bridges built to these units have carried loads, the computed stresses from which are twice these units, without showing any ill effects.

Mr. Seaman shows a 2-track jack-knife draw (Fig. 30). The birth and culmination of this type of draw have been shown[99] in *Engineering News-Record,* by a single-track patent drawing and a photograph of the 8-track draw of the Boston and Maine Railroad Company, over the Charles River channel, at Boston, Mass.

Professor Draffin's notice of the Enoch Hale Bridge is appreciated. It was the first in the United States, that could be called a framed bridge, its predecessors having been of the stringer type, none of greater span than could be negotiated with the magnificent pine spars then available. Mr. Hale secured money from Mr. Geyer, a merchant of Boston, mentioned by Professor Draffin, by a mortgage on his toll bridge. It was a well-paying piece of prop-

[99] *Engineering News-Record,* Vol. 108 (1932), p. 93.

erty which Mr. Geyer had been trying to buy for several years. The mortgage was on short time, and, when its expiration approached, Mr. Hale succeeded in borrowing from his friends to liquidate it. He sent the money to Boston by a messenger who dallied a day at Lowell, and when he tendered the funds to Mr. Geyer, was told that he was a day too late, that the mortgage had expired, that Mr. Geyer had already assumed the title, and that the bridge was his. Such was the law of foreclosure in those days, derived from the Cavalier laws of England, the result being that Enoch Hale lost his noteworthy bridge and became indigent in his old age.

Professor Draffin states in his first paragraph that the present concrete successor of the Hale Bridge was built in 1930. The previous lattice bridge was removed in 1930 on account of failure due to a decayed wall block, but the handsome concrete structure was completed and dedicated with fine ceremony in 1932.

The writers concur in Professor Draffin's conclusion that repairs rather than rebuilding were done by Mr. Geyer when the Enoch Hale Bridge was eighteen years old; and that it was done as window dressing in connection with obtaining a mortgage instead of for strengthening the structure. In fact, there is a book extant containing a picture of the village and landscape made from a point at a considerable distance from the bridge, showing its braced stringer outline with the spandrils filled with lattice trim. Probably this trimming with new fences and floor were the extent of Mr. Geyer's repairs. The statement made by President Dwight about the bridge being an obtuse arch has misled several writers. A braced stringer bridge of that magnitude is comparable to a segmental arch of three voussoirs which make obtuse angles with each other. Fig. 8 is an excellent line drawing of the oil painting made soon after the bridge was erected, but the braces were much more sharply inclined than shown, and the right-hand end of the floor should be raised to continue the grade of the other end to meet the side of the gorge higher than indicated. The highway along the mountain side was much higher than the bridge, so that a steep descent was required.

In speaking of the Tucker Bridge (Fig. 18), Professor Draffin calls it of the lattice timber type. This is technically wrong. It was a single-plank lattice bridge of 3 by 10-in. plank, while Fig. 19 is a timber lattice of 6 by 8-in. timbers. Much more of historical interest could be written of this bridge and its contemporary events than is pertinent.

It is gratifying to read in the discussion by Mr. Morse, his appreciation of the splendid workmanship in the bridges built by Theodore Burr. The writers refer to the same characteristic in those built by Timothy Palmer; and indeed it was commented[20] on by the late Theodore Cooper, M. Am. Soc. C. E.

The writers join Mr. Storrs in honoring the early builders who had little to guide them in planning or building their bridges, except their experience and strong common sense. These men and the early builders named in the paper were real engineers. The historians of America have generally ignored them, perhaps because they were regarded as mere carpentero. They were among the "empire builders." They are more deserving of large mention in history and inscriptions on memorial tablets, than many who have achieved less for their country. They are comparable to the "Brothers of the Bridge" who built and maintained the great stone bridges of Europe during the Middle Ages.

The writers can not entirely subscribe to the enthusiasm expressed by Mr. Hogue in the second paragraph of his discussion, as to a recrudescence of timber construction in the country at large; but in regions of primitive growth, as claimed in the paper, they believe a considerable field is open for wooden bridges. The examples of longevity cited for treated lumber (40 years), seems absurd when applied to truss bridges. If properly covered with roof and side boarding 100 years is frequently attained. An example is cited later in this discussion of one, 120 years old, which is still "going strong."

The ring dowels for timber connections mentioned by Mr. Hogue are certainly worthy of consideration. Astonishing examples of their use are illustrated in recent publications of the National Committee of Wood Utilization, U. S. Department of Commerce, Washington, D. C.

The paragraphs of the main paper describing Fig. 16 answer Mr. Oliver's uncertainty in regard to the writers' conclusions as to the adaptability and economy of wooden bridges. Mr. Oliver does not mean that the Selby Bridge, built in 1790, is still in use, as to its original timbers. Stringers bridges last from 5 to 20 years, depending on latitude, character of lumber, and exposure to the weather, but their members are readily renewed, and hence the bridge, as an adjunct of transportation, lasts perpetually. Preservative treatment will double the durability of their members.

Regarding Pons Sublicius, mentioned by Mr. Oliver, the writers do not think any published picture of its construction is authentic.

The writers are greatly indebted to Professor Laurson for the items relating to early European bridges, which he cites. It is interesting to learn that lattice bridges were built on various railroads in England and Ireland previous to 1846. Ithiel Town's patent for this type of bridge was granted in 1820, and it is on record that he visited England about 1840. No doubt he was successful in selling rights to build by his patent. Regarding "assembling" bridges on land before erection in place: If Professor Laurson means assembling them erect, it is seldom, if ever, done. The only practical way to build lattice bridges and others, such as the Burr and Long types, in which the web members pass through the chord, is to frame the trusses lying flat on skids on shore, or on falsework, if wide enough, and traffic is suspended. Holes for pins and bolts are bored in this position and, when complete, each stick is carried to its place on the falsework and assembled erect. The chords of Howe trusses are framed on the falsework and the trusses erected complete after the lower chords are in place. The practice has been to fabricate and erect these bridges as one job; whereas in metal bridges the two phases of the work are widely separated. This distinction may disappear in the Machine Age, supplemented with new methods of making timber connections;

but it is incredible that the building of wooden bridges should become common in regions where organized fabrication shops are within economic reach of the locus of the bridge.

A wrong impression is conveyed by Professor Kirby's reference to the span of the Schaffhausen Bridge (Fig. 7). It consisted of two spans of 171 and 193 ft. The writers have no data showing the span of the McCalls Ferry Bridge, but mention several of much greater span than the Schaffhausen, including some built previous to 1814. (See Wernwag's "Colossus" of 340-ft span, built in 1812 (Fig. 12), and one at Wittingen, Germany, 390-ft span, built during the third quarter of the Eighteenth Century.)[17] It is an interesting fact that a complete model of the Schaffhausen Bridge is still treasured in the town 134 years after the destruction of the bridge by a French army. This is without the covering and exhibits all the details of construction.[100]

Fig. 32, presented with the discussion by Mr. Thompson, is very interesting. The weather protection mentioned by him can be seen to consist of a kind of flexible shingle applied to each stick of the corbels and cross-ties. The presence of this bridge in far Japan, coupled with the numerous examples of the same type now existing in Northern India, and the Savoy ruin, mentioned in the paper, under "The Ancient Background," seem to prove that the corbelled bridge was known and used across the great Eurasian Continent in the early ages of human development.

Mr. Guppy calls attention to an interesting and true phase of the difference in designing wooden and metal bridges. The comparison might be extended to a somewhat similar difference in fabricating and erecting the two classes. Fig. 23, or its modification, might well be added to Figs. 16 and 28 as examples of the use of metal splices in timber tension members.

The simple method of anchoring a bridge to its abutments described by Mr. Houk, is well worth using on all wooden bridges over streams subject to high water or severe winds. Hundreds of wooden bridges have been floated —pushed by drift or blown by wind—from their bearings, nearly all of which would have escaped serious injury had they been anchored. It seems scarcely necessary to carry the rods to the upper chord. A short rod each side of the lower chord, passing through a yoke block, and each one connecting with two wire lines below, would seem simpler to install than the arrangement described by Mr. Houk.

The combination wood and steel bridges described by Mr. Baker are interesting examples of history repeating itself under new conditions. Mr. Baker's type of construction suggests great possibilities for highway bridges of spans up to 80 ft. Lower chords of channels in one length up to two cars long, with perhaps a plate reinforcement at the center, with angle-blocks and tie-plates riveted on at the shop, or laminated chords of treated plank securely united as one timber with ring dowel, or other connectors, would naturally follow as a type of construction suggested by Mr. Hogue. Preservative treatment is not needed if the ends are protected from dirt accumulations, and if treated

[100] An excellent view of the model appears in Mohringer's "Bridges of the Rhine", pub. in 1931.

wall-blocking is used. Rod verticals and timber upper chords and web diagonals may be used. All spans greater than 45 ft should be housed with roof and side covering, using a roof with a good overhang and a pitch of about 4 in. per ft. Inexpensive trim on facade and sides should be used, and the entire structure painted appropriately to present a bungalow appearance, far surpassing in picturesqueness those of the New England covered type. Shorter spans with lower trusses should have each truss housed, roof and sides. A concrete floor-slab on steel cross floor-beams would prevent the decay complained of, and housing will save the end posts and upper chords. Window openings (see Fig. 13), admit light and increase safety of traffic inside, better the appearance outside, and assist ventilation. The flooring, of whatever character, furnishes adequate lower lateral bracing. The upper chords should be adequately braced in connection with the roof framing.

The writers disagree with Mr. Baker in his statement that wooden bridges are very costly to maintain. It is inferred that the tendency to decay is the cause of his complaint. Piles in sea water, infested with marine insects, open stringers, and plank floors, and, in fact, all timber exposed to the weather, or in contact with masonry or earth, are short-lived in warm climates unless thoroughly treated with a preservative. Truss bridges on the contrary are very durable if protected by housing. (Note the many instances cited in the paper and discussions of bridges a century and more old.) An experience of more than forty years has taught the writers that current maintenance of wooden truss spans is not so costly as maintenance of steel spans. The worst enemy of wooden bridges is fire, due at present largely to cigar and cigarette stubs. Obsolescence and increase of loads, especially in case of railroads, are the usual cause of renewals, rather than physical defects.

Mr. Reed and Messrs. Davis and Jemison have furnished valuable records of early truss bridges west of the Allegheny barrier. These trusses as a whole are based on the Burr design, a worthy example to follow.

The Ramp Creek Bridge built about 1837, with its arches, upper chords, and roof framing of yellow poplar (white wood), and its floor-beams, stringers, and under floor of walnut, causes surprise to an Easterner; but it really shows excellent engineering judgment on the part of its builders and was not less remarkable than the use of the magnificent white pine for bridge work in the East, which timber was once reserved, by law, for masts of the Royal English Navy. The fact that after about 95 years of service the bridge was deemed worthy of re-erection in a location where it would be subject to no heavier traffic than that for which it was built, is certainly worthy of attention.

Failure of floor-beams by longitudinal shear is common if they are overloaded and mortised near their ends for lateral braces, as described for the bridges in Figs. 35 and 37. A well-laid floor is very ample for lateral bracing of a highway bridge. The "bowing" out of line mentioned in the case of the former bridge is quite likely due to flatness of arch and lack of rigidity of the shallow truss with no counters.

The durability of the hand-forged bolts mentioned in the description of this bridge is worthy of note, although not strictly pertinent to a paper on wooden bridges. The more nearly pure an iron is, the better it resists corrosion. Old-time Swedish or Russian iron, or present-day ingot iron, is practically rust proof; while structural steel is not so durable as wood in certain locations.

The Kennedy family, J. J. Daniels, and Lemuel Chenoweth, should be added to the honor roll of names of noted bridge builders.

Another variation of the Burr type of truss was seen in a bridge built by George W. Thayer, of Massachusetts, over the Delaware River, at Narrowsburg, N. Y., in 1859, which had the longest framed trusses (clear span of 262 ft), except Wernwag's "Colossus" (see Fig. 12), of which the writers have knowledge.[101] Each truss had two arches on each side, one above the other, separated by 12 in., or more. A peculiarity in the trusses was horizontal struts ("straining pieces") between the posts, fixed just below the upper, and above the lower, chords where the diagonal braces were framed into the posts. The latter feature was specified in a patent granted to Mr. Thayer in 1845.[102] This and other patents noted in the paper are illustrated and described in a series of articles,[103] by F. B. Brock, Patent Attorney, Washington, D. C., on expired patents for wooden, iron, and combination bridges. The Narrowsburg Bridge was removed when it was about forty years old, while still capable of giving good service (probably for more than twice that period).

The "modern" timber connectors shown by Mr. Landsem are interesting. The multitude and variety of these patented connectors reminds one of the flood of United States patents issued prior to 1880 for bridge trusses of wood, iron, and combinations of the two, each patentee claiming that his arrangement of parts was safer and cheaper than any previously proposed system. The upper chord joint (Fig. 41), seems like a substitute for the American pin-joint. By hooking the four members to a ring in this joint there is no moment induced on the center bolt. Its workmanship must be quite expensive. On this account it is not probable that this joint will be generally used in this country. There is a wide field for their use as smaller connectors in high towers of timber trestles, towers for transmission lines and radio aerials, and especially for portable structures, such as oil-well drilling derricks, observation stands, etc. The old-time rectangular key, with a length of about ten times the depth of dap cut in the timber for it, and two bolts to resist the overturning tendency, is ideal in efficiency, but is perhaps not so economical in these days of machine employment as the circular dap and connector. A simple disk with a central bolt is probably the best solution of the problem.

Mr. Landsem has contributed some valuable facts as to European bridge building, ancient and modern. His statement about the flexibility of laminated arches should not deter builders from their use. If properly secured to stiffening trusses they hold their shape perfectly. Mr. Landsem mentions the materials as being "boards." Arches are built up of 2 to 4-in. "boards" very readily. Radial bolts should be inserted at intervals of not more than 8 ft,

---

[101] *Engineering News*, Vol. 42, p. 122.
[102] *Loc. cit.*, Vol. 9, p. 418.
[103] *Loc. cit.*, Vols. 9 and 10.

and the load applied at points should not be more than 12 ft apart. If greater intervals are allowed buckling may occur.

Professor Finch contributes a fine historic summary—a splendid frame in which wooden bridges occupy but part of the picture. There is little in it for the practical bridge builder to discuss, but the writers feel disposed to comment upon the feature of statical indetermination and the men named as bringing American practice to rational analysis.

It is a fact that the two all-wooden types best adapted for bridges are inherently statically indeterminate; namely, the Burr (Figs. 14 and 15), and the lattice with supplementary arch (Fig. 17). The elasticity of the material, the tolerance of timber joints to slight distortion, and the natural law of gravitation will bring the stress to the most rigid members. Timothy Palmer's Easton bridge (Fig. 11), was a rather indifferent truss yoked to a flat arch for a lower chord. Theodore Burr improved on this, introducing more indeterminacy but also a large amount of practicality, resulting in wide adoption and approval, as shown throughout this paper and its discussion. The lattice with arch is superior to the Burr type in that it requires much simpler forms of timber, much less skilled labor, and its truss is much better adapted to resist reversals of strain. Still the rank indeterminateness is there, and no one but the schoolmen know or care anything about it. If there is strength enough in both members of the combination acting together, the elasticity and safe "give" at joints will distribute the load exactly according to the static rigidity of each.

Regarding the men named by Professor Finch: Colonel Long was a bridge builder of courage and imagination, although his designs were not so extensively copied as were those of Burr, Town, and Howe. Neither Ithiel Town nor William Howe were builders of bridges. Town was a business man who patented the lattice type which was designed and built in Vermont years before the patent was issued. Sanford Granger, mentioned in the paper, was probably connected with Mr. Town, in securing the patent. Mr. Granger, however, did not invent the type. A lattice bridge was built in 1813 (when Granger was only 17 years old) over Otter Creek, in Pittsford, Vt., many miles from his home. The bridge is still in service after 120 years. It was thought worthy of being floated up river to its place in 1927 when the high water of that year carried it a mile or more from its abutments. Town also patented the timber lattice in 1839. This is also of Vermont birth. William Howe was an architect. In 1839, in connection with Amasa Stone, then 22 years old, he built a bridge of 75-ft span for the City of Springfield, Mass., probably on the Howe principle, because in 1840 he secured a patent, and they built the great bridge over the Connecticut River (Fig. 24), as stated in the paper. The next year Stone bought out Howe's rights in the patent and went into bridge building as a business. Later, he formed a company under the name of Stone and Boomer which engaged in railroad and bridge construction on a large scale.

Concerning the reference to Trajan's Bridge over the Danube River by Mr. Eremin, the figures quoted are those given by Nicholas Bergier[16] and Dion

Cassius.[104] The latter was the Roman historian whose notable history of the Roman Empire in eighty books was written between 217 and 229 A. D.—the century following Trajan's death in 117 A. D.[105] The historian's account of this prodigious and unprecedented work states that it was built to facilitate Trajan's second campaign against a revolted enemy, who killed himself after complete defeat in 106 A. D. The same authority puts the building of the bridge about 104 A. D. The brief existence of the bridge is also noted, for the narrator adds that Trajan's successor, the Emperor Hadrian, caused the arches to be torn down because he was extremely jealous of the fame which this 'wonder of the world"—as the narrator styles it—had brought to his predecessor.[106]

The information on foreign bridges contributed by Messrs. Eremin and Andersen is highly appreciated. The double-web systems of the Gulfoss and other bridges of the Howe type are interesting. They admit the use of smaller diagonals and rods than single systems but, for spans such as those mentioned, it is doubtful if the added work of framing and assembling did not more than balance the saving of material. The omission of side boarding seems wasteful. The lower chord needs protection from the elements more than the upper one. The type of trestle shown was formerly used in some sections in the United States and has the advantage, if properly designed, of allowing the removal of any piece of timber for replacement without interrupting traffic.

Northern Russia and Scandinavia with their wealth of coniferous timber growth are the natural homes of the lattice type of bridge. Baltic fir is well adapted for that type if proper units of stress are used. One-inch round iron or steel drift-bolts, instead of the oak pins shown on Fig. 16, would be advisable for connectors.

The suggestion of Mr. Atwood that early obsolescence may overtake a structure built for all time is of much greater moment than many realize. Most engineers have dug out so much that their predecessors built under the slogan, "build for the future"—good, durable, and expensive work no doubt, but ill adapted to present-day needs—that they should feel that their work will probably go the same way, and that they should build simply what is needed now, and pay for it. The other alternative is to do as has been done many times, to wit, build for the future, paying for it with bonds for coming generations to liquidate long after the constructions have been abandoned or dismantled.

Mr. Meyer presents another specimen of the interesting type of corbelled bridges which have appeared at several points in this paper and its discussions. The great heart of the Eastern Continent lying east of the Ural Mountains and including Western China, in which the Kingdom of Muli is situated, has much to reveal of the early history of Aryan peoples. This type of bridge, called "cantilever" by Mr. Meyer, must be accepted as the prototype of con-

---

[104] Appleton's Cyclopedia, Vol. VI, Article on Dion Cassius, p. 122.

[105] *Loc. cit.*, Vol. XV, Article on Trajan, p. 838.

[106] Bergier's "Histoire. des Grandes Routes de l'Empire Romain," Tome 2, livre 4, p. 750.

structed bridges. It is found from the Pacific to the Atlantic Oceans, coming to Europe, apparently, with the Alpine stock of people in very ancient times. Its class had better be called "corbelled," leaving "cantilever" to the distinctive type with the back lever doing duty by spanning an opening.

Professor Howland submits some appropriate information about the engineering literature available in the Eighteenth and early Nineteenth Centuries. It recalls research made for the paper, the results of which were precluded because of limitations of space. The lack of suitable timber in England, and the resort to cast iron by the English engineers was noted in the paper under "Later European Timber Bridges." For example, this appears by consulting the monumental folio atlas accompanying Rickman's "Life of Telford." That portrays, by fine engravings, 22 stone bridges, including masterpieces such as the famous London bridges; 35 road bridges in the Highlands of Scotland—from arched culverts to those of two or more spans—all of stone; 8 bridges built of cast iron, made in sections and bolted together—two over the Thames and others less known; 2 famous suspension bridges; and only 2 timber bridges of minor importance.

Turning to the European Continent a different story comes to light. The treatises on "The Arts of Construction and Carpentry" by French engineers—Gauthey, Bélidor, Bruyére, Cordier, and especially E'my—are prolific in fine drawings of actual or suggested timber bridges in a great variety of arched and braced contrivances. Perronet, who made his great stone bridges patterns (as it were) for English engineers and the world to follow, developed the designing—as Professor Howland has described—of the heavy timber centers which he required. It is in order to add that the splendid atlas of plates portraying his works has only one drawing of a timber bridge, and that was a "project" for a bridge over the Seine, in Paris, with seven arches of 60-ft span and 10-ft rise, on piers of timber piles. Evidently, he preferred to build for posterity.

Mr. Ellis corroborates others in showing the general prevalence of the Burr type of bridges in certain sections of the country. In some others, particularly in Northwestern New England, the Town lattice prevailed. Burr's designs were not conspicuous for scientific proportioning of parts. His experience and natural flair for heavy construction were probably his only guides. The later changes in character of highways and of loads carried, would find weak spots in many of his bridges, especially those built by others who had not his flair and who built at his type rather than to it. The combination truss and arch, which is the real essence of the Burr patent as finally adjusted, is the absolute final acme of design in wooden bridges of considerable span.

Mr. Ellis adds cumulative evidence to the value of effective covering which is stressed strongly in the main paper.

Mr. Bell's discussion opens an interesting line of conjecture. He states that the bridge at Camp Nelson, near Lexington, Ky., was built by Lewis Wernwag in 1838, and he feels certain that the one at Cheat River (Fig. 35), was constructed by the same builder. In 1855, Nicholas Powers built the

North Blenheim Bridge (Fig. 20), using the same principle of construction—an arch of three rings on the center line of the truss—but he carried his middle truss to the ridge pole of the pitched roof, which brilliant idea, was missed by Wernwag, both at Cheat River and at Camp Nelson. Powers adopted the later type of Long's patent truss which is better adapted for stiffening the arch than the Burr type. The arches of these three bridges follow those of Wernwag's famous "Colossus" in design, but in these later designs he adopted the parallel chord idea of Burr for the stiffening truss instead of framing it on top of the arch (see Fig. 12), thereby giving better support for floor and roof than the original example.

Without doubt the first example of an arch on the center line of a truss was at the Cheat River Bridge, and its author was Lewis Wernwag, who died at Harpers Ferry, Va., in 1843. The two examples show that he adopted, as his later practice, the arch as the main carrier of loads, placed on the center line of the truss designed to hold it from distortion under concentrated loads, even to the extent of the twin arches embedded in a single truss, as at Camp Nelson. His arch segments were hewn to the curve of the arch. This involves the weak factor of timber (longitudinal shear) to a slight extent; and laminated arches laid up with planks 3 to 4 in. thick are now generally used. It remains to examine the truss factor to determine the best type to use for the heavy concentrated loads of the present time both for highway and railroad service. Manifestly, a truss, all the members, chords, and web of which will act in tension or compression, is ideal for the purpose. The Town lattice fills these requirements perfectly, and hence must be given preference. It is gratifying that the idea can be traced back to the illustrious Wernwag.

The sentiment expressed so strongly by Mr. Bell that wooden bridges should not be scrapped because of obsolescence and possibly weak floors, is heartily seconded and upheld by the writers.

Considerable has been written in these discussions about preservative treatment of timber for various purposes connected with bridges. In other publications, treated timber for heavy truss bridges of long span, has been advocated so that roofs and side covering would not be needed. The advisability of the use of preservative to this extent has not been proved either by the discussions or by the experience of the writers. For trestles, small deck spans, and short spans unsuited for covering, treated timber is advisable; but for trusses of spans requiring joints in lower chords, covering is demanded. Esthetic reasons, as well as economy, require covering of trussed spans. What looks worse in a landscape than an uncovered wooden through truss bridge especially if it is black with preservative? The bridge shown in Fig. 29, for example, was rightly left uncovered for temporary use only; but its unfinished aspect shows how those designed to be permanent would look.

The general trend of these discussions amply proves the longevity of wooden bridges and justifies their use in localities where they are substantially cheaper than bridges of metal or masonry.

A chronology of the development of the types described in this history and its discussion is given herewith; the first twenty-three types, in this list, built during a period of nearly fifty years, were framed by use of treenails and such small hardware as could be forged by blacksmiths:

Pile-trestle and stringer bridges, crib and stringer bridges after European antecedents.............	From 1620 forward
Enoch Hale's braced-stringer bridge, at Bellows Falls, Vt....................................	1785
Timothy Palmer's Essex Merrimac Bridge........	1792
Timothy Palmer's Piscataqua and Haverhill Bridges	1794
Timothy Palmer's Georgetown, D. C., Bridge.....	1796
Timothy Palmer's "Permanent" Bridge, at Philadelphia, Pa	1804–06
Timothy Palmer's Easton, Pa., Bridge............	1805–06
Graves' (second) Connecticut River Bridge, at Hanover, N. H..................................	1796
Windsor, Vt., Bridge, contemporary with the Graves' Bridge	1796
Theodore Burr's Waterford, N. Y., Bridge........	1804
Theodore Burr's Trenton, N. J., Bridge..........	1806
Theodore Burr's Mohawk River Bridge..........	1808
Theodore Burr's Harrisburg, Pa., Bridge.........	1816
Lewis Wernwag's "Colossus" at Philadelphia, Pa..	1812
Lewis Wernwag's New Hope, Pa., Bridge........	1813–14
Lewis Wernwag's "Economy" Bridge	1810
Earliest lattice-truss bridge of which there is record	1813
Ithiel Town's plank lattice truss, patented........	1820
Truss of Stephen H. Long, patented.............	1830
Ithiel Town's timber lattice truss, patented........	1839
Wernwag's Cheat River Bridge, West Virginia....	1834
Wernwag's Camp Nelson Bridge, near Lexington, Ky.	1838
(Standing in 1933 after 95 years. In both these bridges the arch is on the center line of the truss.)	
The Ramp Creek Bridge, Indiana, Burr trusses, built in	1837
(Renovated and in service, 1933, after 96 years.)	
The Raccoon Creek Bridge, Indiana, Burr trusses.	1838
(Still in use, 1933, after 95 years.)	
Brunel's experiments in England with preservatives, kyanizing, etc...........................	1835
Wooden lattice bridges on British railways after Ithiel Town's visit about 1840, before..........	1846
William Howe's patent for the Howe truss........	1840
William Howe's Connecticut River Bridge, at Springfield, Mass.	1840

The Tucker Bridge, at Bellows Falls, Vt., plank lattice	1840
The trusses of Thomas W. and Caleb Pratt, patented	1844
Typical Burr truss railroad bridge (framed with white pine), at White River Junction, Vt.....	1848
Howe truss bridge with double arches, at Bellows Falls, Vt.	1850
The unclassified truss of Nicholas Powers, North Blenheim, N. Y.	1855
First bridge across the Mississippi River, five spans, Howe trusses with double arches, at Rock Island, Ill.	1853–56
Second historic bridge at Rock Island, Ill., Howe trusses with curved upper chords and no arches, some time before	1868
The Ledyard Bridge, at Hanover, N. H., timber lattice	1859
Howe truss bridge with double arches (12 spans) at Havre de Grace, Md.	1862–66
The gallows-frame drawbridge	1660
The **A**-frame drawbridge	1835
The jack-knife drawbridge	1845

Within less than a month after publication in "Proceedings" of the closure of the discussion on "A History of the Development of Wooden Bridges," Mr. Snow, joint author of the paper with Professor Fletcher, died suddenly on September 4, 1933, in his eighty-fifth year. His associate in the research can do no less than testify that whatever of authoritative value this paper may possess, is due in very large measure to Mr. Snow's part in its production. Mr. Snow's long experience as Bridge Engineer and, later, as Chief Engineer of the Boston and Maine Railroad—covering the period when wrought iron gradually superseded timber in bridge construction—made him an outstanding authority on the subject. His knowledge of details and his experience in maintenance of such bridges constrained him to adopt a conservative policy in retaining timber bridges as long as they could be made assuredly safe under increasing loadings.

RAY E. WILSON

Among the greatest American covered bridge ''buffs'' is Ray Wilson who has personally visited over 1300 of these structures in North America and Europe. As an industrial engineer (Massachusetts Institute of Technology, 1912) he specialized in protecting industrial plants from fire and explosion. From before World War I until after World War II, Mr. Wilson has worked with many private firms and government agencies in pursuing his specialty. His work was particularly valuable in wartime when the threat of sabotage was great.

Today, Mr. Wilson devotes much of his time to writing and lecturing on all aspects of covered bridges, but particularly ''their preservation as an American Heritage''. Among his most popular writings is the one printed here which first appeared in his alma mater's ''Technology Review'' in June, 1971.

A Glossary of Terms Applying to Covered Bridge Trusses

Brace: A diagonal timber which slants upward toward the center of the bridge.

Chord: The horizontal timber or timbers forming part of the truss; a lower chord supports the floor, an upper chord the roof.

Counterbrace: A diagonal timber or rod which slants upward away from the center of the bridge.

Deck: The floor or roadbed of the bridge.

Diagonal: A timber or rod which slants, usually between two panel posts.

Flared post: A post with the top slanting toward the end of the bridge.

Footpiece: A specially designed casting, usually placed between diagonals and chords for rigid assembly of members.

Laminated arch: A series of planks bolted together with staggered joints to form an arch; used in lieu of a solid timber.

Panel: That portion of a truss included between two vertical posts.

Post: A vertical or upright timber in a truss.

Span or clear span: The length of a bridge between abutments or piers. A truss span is the length between the end posts of the truss.

Truss: An arrangement of timbers and rods to support each other—as well as other loads—rigidly. This requires the use of a suitable combination of triangles.

Anticipating readers' summer vacations, an ardent "covered bridger" provides an engineer's guide to the different structural forms which can be seen in the 990 remaining "bridges to America's past"

Raymond E. Wilson

Reprinted from: *Technology Review,* Massachusetts Institute of Technology, May 1971.

Twenty Different Ways to Build a Covered Bridge

Though covered bridges form a colorful link to America's past, readers of *Technology Review* may be surprised to learn that this structural form is neither unique to the New World nor original in it. Indeed, Marco Polo recorded Chinese bridges "with very handsome roofs" in journals of his travels 800 years ago; and there are 230 covered bridges still in use in Austria, Switzerland, and Germany. The oldest in Europe, and probably in the world, is the Chapel Bridge in Lucerne, a footbridge built in 1333.

Most European covered bridges are of the timber trestle type, using short spans set on stone piers or wooden supports. A few of the older and larger Swiss bridges use massive arches with heavy radial struts and braces. The truss—though a sixteenth-century Italian invention—was not used in Europe until after it became common in America.

The basic form of the truss is the triangle, the only geometrical figure that cannot be distorted under load. All truss bridges, therefore, are designed with a single triangle or a series of adjoining ones. Andrea Palladio, an Italian architect, became a leader in the design of trusses and in 1570 published a book in which he showed four types—a queenpost design, a multiple kingpost truss, and two panel trusses with arched chords and X-diagonals. Palladio actually built one or more bridges incorporating such truss designs, and his treatise was published in English in 1742. But he was far ahead of his time, and his work was subsequently neglected by European builders, who favored the stone arch.

Only 200 years later, when bridge builders were confronted with the urgent problem of spanning large rivers of the North American continent surrounded by plentiful timber, was the truss rediscovered, and Palladio's sketches were undoubtedly studied by the American carpenters of the eighteenth and nineteenth centuries, who were skilled in the building of large barns and farmhouses.

Preserving the Structure

The first bridge of any appreciable length in the United States was a braced arch structure built by Enoch Hale in 1785 across the Connecticut River at Bellows Falls, Vt. Many New England carpenters thereafter experimented with bridge construction and developed a braced arch design, of which a number were built over the larger rivers. Finally, in 1805, the first covered bridge was built by Timothy Palmer, "bridge builder extraordinary" of New England. This was a three-span, 550-ft. structure using an arch truss with an arched roadway over the Schuylkill River on Market Street, Philadelphia. It is recorded that the bridge committee insisted that the heavy timbers be covered with a roof and siding as protection against the weather—the sole purpose, incidentally, of the bridge enclosures. The life expectancy of an open bridge was seldom over 10 to 15 years due to rotting of the timbers, while covered bridges have been maintained over 100 years.

After completing the Permanent Bridge in Philadelphia, Palmer built but one other structure—crossing the Delaware River in Easton, Pa.—before retiring. However, the point had been made, the young country was developing rapidly, and there was tremendous demand for covered bridges to span the many large rivers along the Atlantic coast. Louis Wernwag, a German immigrant with demonstrated skill as mechanic and bridge builder, realized the situation and in 1812 secured the contract to build the Colossus

Bridge, a 340-ft. structure across the Schuylkill, which he designed and successfully completed as an arch truss. His reputation established, he erected some 30 other large covered structures between 1810 and 1838, using heavy arches with flared multiple kingpost trusses on either side. None of these bridges now remains, although the occasional use of flared posts in other types of designs causes some confusion as to their proper designation.

Theodore Burr, the third and greatest member of the pioneer group of covered-bridge builders, was born in Connecticut in 1771, moved to New York to become a skilled carpenter-mechanic and operate his own saw mill, and finally located in Harrisburg, Pa. He, too, had adopted Palladio and designed a large bridge using a multiple kingpost truss, the first to depend upon a truss without an arch. But he found quickly enough that such a bridge was not stable under live loads, and he was forced to strengthen it with arches. On this basis he continued to build large bridges throughout the Northeast—including a successful 400-ft. span across the Hudson River at Waterford, N.Y., in 1804 using an X-panel truss and two arches; this bridge—probably the first arch bridge to have a level roadway—was not covered until 1814.

It was in 1817 that Burr patented his revised arch-truss design using a multiple kingpost truss which soon became the most popular covered bridge structural system. Unfortunately, he soon thereafter extended his activities too far, accepting contracts for five huge structures over the Susquehanna River at once. He failed financially, and this was undoubtedly a factor contributing to his early death in 1822.

These are among the highlights in the early history of a romantic chapter in American engineering which has now attracted a company of dedicated hobbyists. It has been estimated that over 10,000 covered bridges were built in the U.S. between 1805 and 1885. All of the truly massive old structures—ranging in length from 500 ft. to the record 5,690 ft. in Columbia, Pa.—have long since been replaced by steel or concrete bridges. But in 1955, 1,564 covered bridges remained; by 1970 the count was down to 990. The longest still standing is 468 ft. long, over the Connecticut River at Windsor, Vt., of Town truss design, built in 1866. The oldest, near Lewisburg, Pa., of Burr truss design, was built in 1812. Canada has over 400 covered bridges, mostly in Quebec and New Brunswick and mostly of Town and Howe truss design, respectively; only a few of these were built before 1900, and they, too, are rapidly disappearing. The longest is in Hartland, N.B.—1,282 ft.—and there are four others which measure over 500 ft.

"Kissing Bridges"—and Other Uses

The most common reason for the loss of covered bridges today is local apathy, with resulting lack of maintenance and collapse. Floods and windstorms continue to exact a toll, but in recent years arson has become a more serious problem; some 70 covered bridges have been so destroyed in the past ten years. Careless truck drivers with too-heavy loads at excessive speeds are responsible for losses, and the increasing number of very large reservoirs for flood control and recreation have contributed to the toll.

Covered bridges once played a vital—and unique—part in the life of nearby communities. Their importance as communication links was fundamental, and is obvious. But the bridges were often the largest covered areas available to the communities, and so they were used for political rallies, revival meetings, and even for weddings. Many a courtship was fostered in these "kissing bridges"—to the delight of youngsters hidden in the rafters. They served as welcome shelters in sudden storms and as protected sites for robberies and murders. The inside walls and timbers were used for communications, carrying advertisements of local stores, circuses, patent medicines for man and beast, and even religious mottos; some of these—as well as many carved initials and hearts—are still legible.

During the past 11 years this writer has visited nearly all of the covered bridges in the United States, about one-third of those in Canada, and a number in Europe. When we first became interested in this hobby, the features which impressed us most were the huge arches with the heavy timber trusses between, which are the marks of the Burr truss. We soon discovered 75 of these bridges within 100 miles of Swarthmore, Pa., where we live. With an engineering background, we quickly became fascinated in studying the construction details and marvelled at the superb hand workmanship of the builders in hewing and assembling the heavy timbers.

As the range of our "bridging" travel was extended to New England, the South, and the Midwest—particularly to Ohio and Indiana—we found covered bridges of many types. Eventually we have come to identify 21 different structural forms, not including variations. Except for information published in two books by Richard S. Allen, we were forced to gain our knowledge of many of these structural variations and their modifications the hard way.

Having accumulated such data, we can now list below the different types of trusses and arches which may be found in the 990 existing U.S. covered bridges—and one in Canada. Readers inspired by this list to seek and study covered bridges during their summer vacations will find that their quest carries them off the speedways to quiet country roads and small villages, that it opens up new friendships, and that it is a uniquely rewarding way to enjoy an auto trip. Perhaps they will become—like the author—"covered bridgers."

Suggested Reference:

World Guide to Covered Bridges. South Peabody, Mass.: National Society for the Preservation of Covered Bridges, Inc. ($2.50).

Raymond E. Wilson graduated in mechanical engineering from M.I.T. in 1912, and for most of his career thereafter he held engineering and management assignments with the Associated Factory Mutual Fire Insurance Co. in Providence, Boston, Philadelphia, and Washington. He has made Swarthmore, Pa., the headquarters for his covered bridge excursions throughout the North American continent since his retirement in 1963.

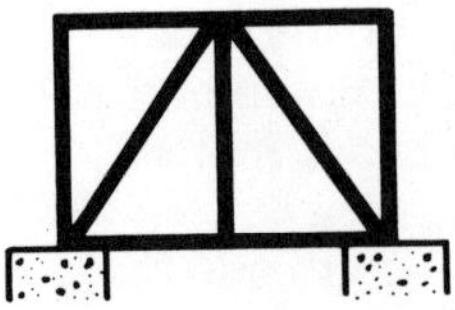

1. The Kingpost Truss. The kingpost is the oldest and simplest truss, used in roof building for hundreds of years. It consists essentially of the basic truss triangle with two timbers slanting down from the center to the ends of the lower chord of the bridge. The kingpost extends down vertically from the center to the lower chord, forming two combined triangles. This post is a tension member, with the diagonals and lower chord in compression. There are 53 covered bridges of this type in use, half of them in Pennsylvania. These bridges are seldom over 30 to 35 ft. long, but with the use of reinforcing timbers and rods—diagonal, horizontal or vertical—spans of over double this length have occasionally been noted.

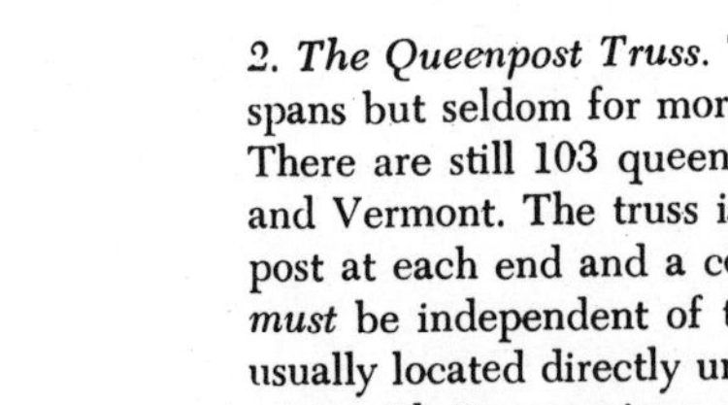

2. The Queenpost Truss. The queenpost can be used for longer spans but seldom for more than 60 to 70 ft., even when braced. There are still 103 queenpost bridges, mostly in Pennsylvania and Vermont. The truss is built as a truncated triangle with a post at each end and a connecting horizontal crosspiece which *must* be independent of the top chord of the bridge and is usually located directly under it. A variation is the pony queenpost, with its crosspiece perhaps only 5 to 10 ft. high. The truss is usually braced in the open, center section, using various arrangements of timbers and rods.

3. The Burr Truss. The Burr Arch Truss is commonly used for longer spans, 100 ft. or more, although a few are but 50 ft. in length. There are nearly 300 still standing, of which 175 are located in Pennsylvania, 70 in Indiana, 15 in Ohio, and 17 in New England. The truss is of the multiple kingpost type with panel posts generally spaced 10 ft. apart. Between each

panel post there is a diagonal timber brace, sloping up toward the center of the bridge. A heavy timber arch is fastened to each side of the truss; for the longer bridges there are double arches. The arches vary in height from about 4 ft. to perhaps 12 ft., or to the top chord. The posts are normally vertical but are sometimes flared slightly with tops slanted towards the end of the bridge; in a few bridges the posts are set nearly at right angles to the arch. Often, as in Lancaster County, Pa., vertical suspension rods, running down from the arch to the lower chord and equipped with screw ends, nuts, washers, and blocks, have been installed in each panel. With this arrangement an appreciable part of the live load is transferred from the truss to the arches. In the older bridges, arches are usually of heavy hewn timbers, but laminated arch construction—up to 12-ply, generally of 2-in. planks nailed together with staggered joints—is found in the newer structures. Some of the first trusses built by Burr had a diagonal-X panel, but only one or two can now be found and these are not too old.

4. The Multiple Kingpost Truss. This is the truss used in the Burr Truss and described above. There are 101 bridges with this type of truss, 82 of which are in Ohio; most of the latter are 50 to 100 ft. long. There are many modifications, variations, and reinforcements of the basic design. The bridges were obviously built by local men; hardly any two are alike, and this fact makes possible an interesting study, found only in Ohio. One unique feature is the open center panel, often found when there is an odd number of panels. An X center panel is occasionally found in others. Another unique structure is the skew bridge, set at an angle to reduce or eliminate the angle of entrance; horizontal, vertical, and diagonal tie rods are used for reinforcement as well as a special U-guy arrangement.

Others have the braces reinforced with half-counterbraces. There are a number of composite designs, some combined with rod counterbracing, like a Childs Truss (*see below*), others with outside and inside guy rods, and others with some Long Truss X-panels (*see below*). Seven bridges have single inside arches and queenposts. There are three bridges which have cambered or curved flooring. The bridges in other states are of orthodox design.

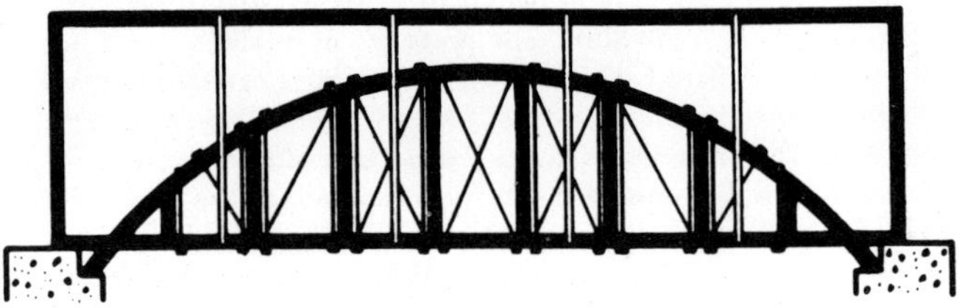

5. *The Arch.* The wooden arch has occasionally been used in lieu of a truss, usually tied at the ends to the lower chord of the bridge. The live load is carried by iron suspension rods running from the arch to the lower chord, using blocks, washers, and nuts for adjustment. Vertical timbers are placed close to each of the suspension rods, and there may be auxiliary diagonal suspension rods. There are seven examples of this type—four in Vermont (one in East Charlotte, two near Brownsville, and the Lincoln Bridge in Woodstock). The vertical suspension rods in the Lincoln Bridge are supplemented by a system of crossed diagonals not unlike a Pratt design. There are two crudely built covered bridges in Newport, Va., with straight timbers set to form an arch between the panel points and many small wooden posts between panels, as well as suspension rods. The seventh bridge, in East Minford, Sciota County, Ohio, has two diagonal timbers instead of an arch but with vertical suspension rods to carry the load.

6. The Town Truss. The Town Truss was designed and patented in 1820 by Ithiel Town, a New Haven architect, who realized the need for a covered bridge truss which could be quickly built by any good carpenter. His bridge was the first truly American design and consisted of a web of light planks criss-crossed at an angle of 45° to 60° like a lattice and fastened together with wooden pins or trunnels at each intersection. It was light and cheap and could be assembled in a few days' time. The design was well received, and—thanks to good advertising and many scattered agents—it soon became most popular both for highways and later for railroads, in which case it was more heavily built and reinforced with extra chords. Some 165 covered bridges of this type may still be located, some with an auxiliary inside arch, particularly in New England, Pennsylvania, Ohio, and the South.

7. The Long Truss. The Long Truss was designed and patented by Colonel Stephen H. Long, an Army engineer of Hopkinton, N.H., in 1830, at the beginning of the railroad era. The truss was of the panel type with crossed single timbers between wooden posts. Long, like Town, was a promoter and competed successfully in selling his truss for highway and railroad use from Maine to the Deep South. The truss was essentially a multiple kingpost with counterbraces. Although the patent showed two single diagonals, practically all bridges now in use are built with two braces and one counterbrace in each panel, and with either one or two posts. The arrival and immediate popularity of the Howe Truss in 1840 affected the number of Long Trusses built, and there are today only 19 left, most of them well over 150 ft. in length. Among these is the well-known double-barrel Blenheim Bridge in New York and Rowell's Bridge in New Hampshire. Both have arches, and the latter has double-intersecting diagonals, the only example extant. Another interesting modified Long Truss which is combined with a Childs Truss and has an auxiliary arch is located at

Odaville, W. Va. Oxford, Ohio, also has an interesting Long and Childs combination. There is a unique "home-made" design in Thomas Store, Ga.

8. *The Paddleford Truss.* The unpatented Paddleford Truss is only found in New England, where there are 23 examples still standing, mostly in New Hampshire. The truss was designed by Peter Paddleford of Littleton, N.H., who had previously erected several fine Long Truss bridges. About 1846, he remodelled this design by replacing the counterbraces with a stiffening member fastened to the inside of the posts at points near the tops and bottoms and extending to the top and bottom chords. This resulted in an unusually strong and rigid structure. Other local builders became interested, and a considerable number were erected. Twelve of these have a heavy inside arch, probably making them the strongest of any existing type of truss.

Howe (usual)

9. *The Howe Truss.* In 1840, William Howe, who came from a family of inventors in Massachusetts, conceived and patented a truss similar to the Long but with a most important improvement. He substituted iron rods for the wooden post as tension members, eliminating one heavy timber and providing a means of easy adjustment by having screw ends with washers and nuts. The rods could be easily shipped and the truss timbers prefabricated, making possible a quick assembly. This truss gradually replaced other trusses; in fact very few covered bridges

Howe (single)

Howe (Western)

other than the Howe were erected after 1880 and it was easily the most popular design during the last half of the century, in the East, Middle West and on the West Coast, where many new covered bridges appeared well into the 1900's. Like the Long Truss, the patents showed single timbers for posts and diagonals. Actually most bridges in the East and Midwest have two diagonal braces and one counterbrace in each panel and two vertical rods between, except that often the end panels have three rods. There are but five Eastern bridges not so designed. These are located at Conway, Mass., Bean Blossom, Ind., Marysville, Va., and two in South Carolina. We may designate these as "single Howe," as they have but one diagonal brace and one or more panel rods. Among the newer bridges of Oregon there are many single Howes, and some counties adopted this as a standard type. Here there are usually a pair or more of diagonal braces and two or more panel rods. The center panel often has crossed diagonals, making a most pleasing appearance with partially sheathed sides. Fifty of the 162 Howe bridges are in Oregon, 30 in Indiana, and 20 in Ohio.

10. The Haupt Truss. The unusual Haupt Truss design was patented in 1839, and two modified examples are still standing. General Haupt, a Pennsylvanian, was a noted bridge engineer and designer who graduated from West Point and became Chief of Military Railroads during the Civil War. His patents show a panel-type truss using single-latticed diagonal braces, each spanning three short panels and also braced with a full-length kingpost, serving in place of an arch. A number of bridges were built in different sections of the country, but only two remain. One of these is in Thetford, Vt., and has 4-ft. panels. A set of single outside braces each overlap two panel posts; there is an inside arch. The other bridge is in a park at Claremont, N.C., where double diagonal braces are placed at either side of the posts to span only two 4-ft. panels. There is no arch.

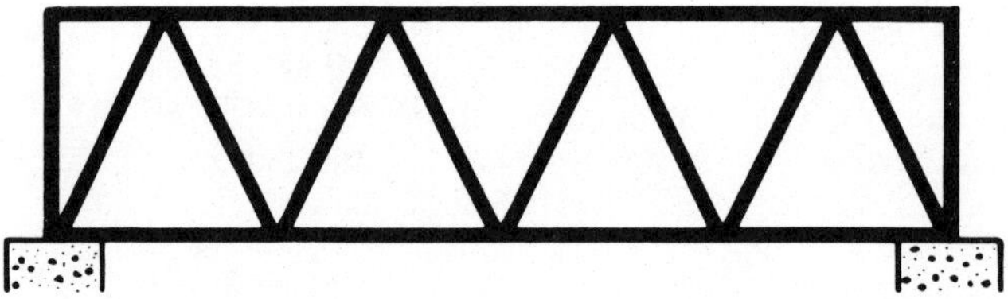

11. The Warren Truss. A simple light-weight truss was patented in 1838 by two Englishmen, James Warren and T. W. Morzani. It is a well-known steel bridge design, particularly for railroads, and is still so used both in England and the United States; it was used to some extent as a timber bridge in the Middle West and later on the West Coast, due to easy assembly and maintenance. The Warren Truss is built in both the single and double system types; the single type consists of a series of diagonal timbers placed in the form of a W with no rods or panel posts. The double system has a second similar set of diagonals which intersect the first. There are still three covered bridges of the single type in Ohio, two having arches added. One without arch is in Muskingum County, owned by the Southern Ohio Covered Bridge Association who use it for meetings and picnics. A double-system Warren Truss, built with double timbers, may be found in Greenup County, Ky., where it is known as the Bennett's Mill Bridge. There is also a single Warren Truss, built with double timbers, in the nearby Oldtown Bridge.

12. The Pratt Truss. The Pratt Truss was the invention of T. Willis Pratt of Boston, an architect and engineer specializing in railroading. He secured a patent in 1844 for a panel-type truss using wooden posts and crossed diagonal rods, just the reverse of the Howe Truss. As designed it proved difficult to adjust; so the "braces" which were tension members were removed leaving only the diagonals which sloped up towards the end of the bridge. As an all-steel truss it was adopted by railroad and highway engineers as a standard type which is still used throughout the country. There is such an all-steel truss in the covered bridge at Keener, Ala., with latticed posts and two-rod diagonals. There are also two examples of the revised wood and steel design in California at Honey Run and Felton.

The Teco Truss. The World Guide lists five covered bridges as having Teco Trusses, which we understand were designed by the Teco Engineering Co. of Washington, D.C. All are of prefabricated timber and are allegedly fitted with special metal connectors or footpieces at the chords for added rigidity. These bridges vary in age from 5 to 20 years. The oldest, in Greenfield, Mass., has a double-intersecting Long Truss with four diagonals and a single post in each panel. The three in Sheffield, Mass., Pepperell, Mass., and Champagne County, Ill., are of the Pratt type with heavy three-timbered diagonals and timber posts. The Bedford, N.H., motel footbridge has a single-timber Pratt-type truss. The County Bridge in Hancock, N.H., listed as "part steel," has inner sheathing, preventing inspection. This 80-ft. bridge was built by Hagen-Thibodeau, Wolfeboro, N.H., in 1937, and has a six-panel, all-timber Pratt Truss with double timber diagonals and chords and single posts, using Teco special split-ring connectors.

13. The Childs Truss. The Childs Truss was patented in 1846 by Horatio Childs of Henniker, N.H., one of three brothers who were subagents for the Long Truss and had for many years been building covered bridges of the Long design for railroads and highways in New Hampshire and adjoining states. The design is basically a Long Truss with counterbraces replaced with a tension rod, resulting in a lighter adjustable truss. Strangely, the design was not used in the East. However, Everett Sherman, a New Hampshire man who had moved to Ohio and was probably familiar with the patent, built a number of these structures in the 1880's. There are now seven of these in Preble County and one in Delaware County. They are still in fine condition, undoubtedly due to the maintenance care of Seth Schlotterback, County Engineer, who is a devoted bridge fan, recently retired.

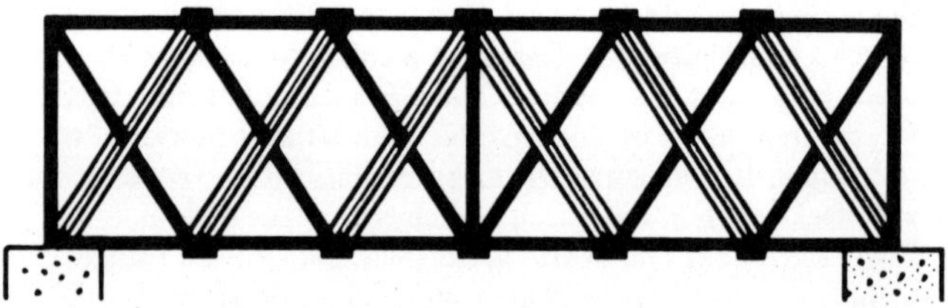

14. The Brown Truss. The Brown Truss was patented in 1857 by Josiah Brown, Jr., of Buffalo, N.Y. It is a very light and simple design built of single intersecting diagonals set at 60°. Their ends are mortised and bolted together with the upper and lower chords to provide rigid joints. There is a center post but no end braces or vertical tension rods. It somewhat resembles a Howe Truss though it is easily distinguished by the lack of panel rods. This bridge truss may be found only in the three structures located in Kent and Ionia Counties, Michigan.

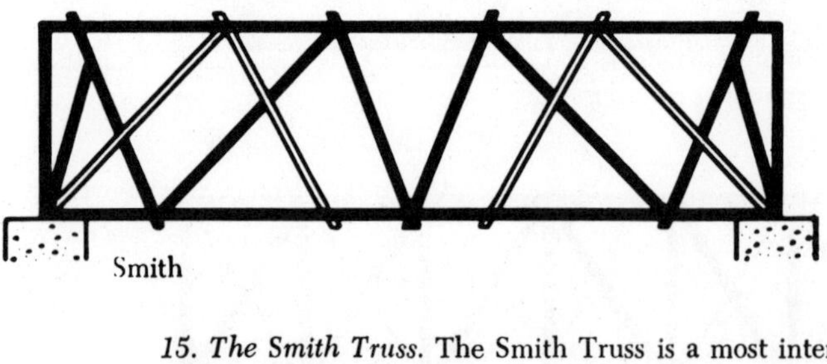

Smith

15. The Smith Truss. The Smith Truss is a most interesting form of light but strong construction. It was designed by Robert J. Smith of Ohio, a successful inventor, builder, contractor, and

promoter, who often provided prefabricated structures built at his mill in Tippecanoe City, which he later moved to Toledo. He patented his first truss in 1867 and a second in 1869, and he later developed—but did not patent—reinforced (type 3) and further improved (type 4) versions. The 1867 type was probably never used. There are 27 covered bridges of the three other designs located in Ohio, Indiana, and (one each) Mercer County, Pa., and Santa Cruz, Calif. Type 2 is built of single timber diagonals with counterbraces set at 60° and bolted to the upper and lower chords. The intersecting braces, set at 45°, are arranged with ends against the counterbraces. The center panel has an open V. The end diagonals are braced and the roof framing is specially designed for rigidity. The reinforced type 3 is similar to type 2 except that two braces have been added to the open-V center panel. The improved type 4 is of substantially different design and was not used till 1875. It consists of double diagonals the full length of the bridge, with a second set of single diagonals intersecting the first. All are set at 60°, and the end diagonals are braced.

16. The Partridge Truss. In 1855, Reuben L. Partridge, a carpenter of Marysville, Ohio, designed and patented his first covered bridge which he termed "self-supporting." It became successful, and it is claimed that he built over 100 in Union County before 1872, when he obtained a patent on his "high truss" improvement and continued his successful career. The design is remarkably similar to that of Robert Smith, showing a series of single intersecting diagonals with 60° counterbraces bolted to the chords and with 45° braces between as shown in the Smith design. A special "bifurcated shoe" or metal footpiece is added to fasten the ends of the braces at the chords, and there were also included braces for the end diagonals. There are presently five of these bridges of a much modified type in Union County and one in Franklin County. None follows the patent design closely, and no two are alike. One to three timbers are used for braces and one or two for counterbraces. Most have been reinforced liberally with diagonal and vertical rods. All are striking in appearance and well maintained.

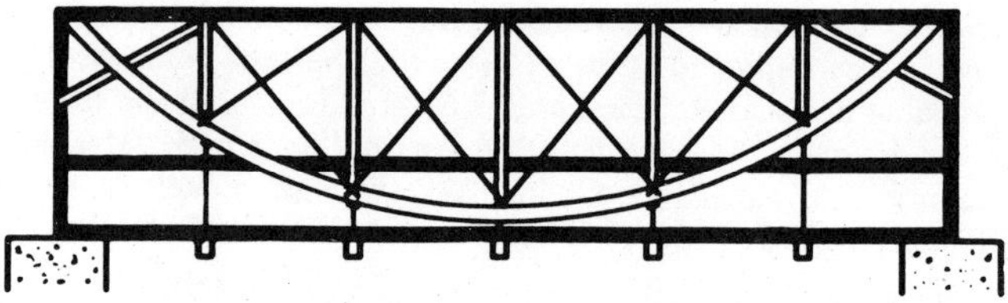

17. *The Bowstring.* Ohio, with its many types of covered bridges, provides a real treat for covered bridge fans with two examples of a bowstring suspension-arch covered bridge, one in Lancaster County and one in Montgomery County. The Jonathan Bright Bridge, 2½ miles southwest of Baltimore, has an 80-ft. solid-timber tied arch to which the five posts are fastened. Double suspension cables are hung from the ends of the upper chord, running through the arch and down to the deck level at the center. In addition there is a system of adjustable crossed iron ties which run from the top chord to the suspension cables. The bridge was built by the Hocking Valley Bridge Co. of Lancaster in 1881. Nearby there is an open iron bridge of similar construction except for the arch. The second covered bridge, built in 1870 in Germantown, Ohio, is 103 ft. long. Bowstrings or flat iron suspension members are swung to the floor level from each end of the upper chord with crossed tie rods, also running down to the bowstring in each panel. There are special U-shaped stirrups from the bowstring to the deck beams. The sides of the bridge are open. The bridge was moved to its present site in 1911.

18. The Post Truss. The single example of the Post Truss in the world is the Bell's Ford Bridge (two spans, 325 ft. long) in Seymour, Ind., which has recently been by-passed. The truss was patented in 1863 by Simeon S. Post, a New Hampshire inventor and railroad engineer. It is of most unusual design, of panel construction. The wooden posts slope slightly inward and there are double-rod braces. The double-rod counterbraces span two panels. It is hoped that local maintenance will be arranged and a park provided for this historical landmark, built in 1869.

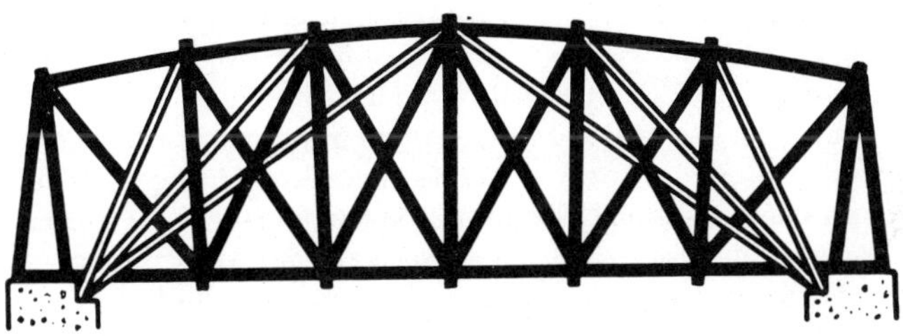

19. The McCallum Truss. Although there are no longer any covered bridges of this type in the United States, there is still one McCallum Truss standing in Quebec, just north of the village of Chateauguay, N.Y. The truss was patented in 1851 by D. J. McCallum, who had become one of the best-known railroad bridge builders in the country. He called it "McCallum's Inflexible Arched Truss," indicating a feature valuable to railroads. It is essentially a Long Truss with timber braces extending from the abutments to the tops of three end panels. It also had an arched top chord. This was a very satisfactory structure, and—despite the difficulty of erection without a skilled crew—at least 150 of these bridges were once in use.

LLEWELLYN N. EDWARDS

Llewellyn N. Edwards was born in Otisfield, Maine on October 6, 1873. The University of Maine granted him a B.C.E. in 1898 and his professional degree, C.E., in 1901. His early career was spent as a structural draftsman in Boston, Chicago, Pittsburgh and elsewhere. He gradually specialized in bridge structures and in 1912 Edwards received his first major assignment; engineer-in-charge of the Coteau Bridge over the St. Lawrence River at Valleyfield, Quebec. In the Great War he served as a captain of engineers and afterwards joined the Bureau of Public Roads as a bridge specialist. His region included Arkansas, Louisiana, Oklahoma and Texas. He left the federal government to return to his native Maine and served as the state's bridge engineer from 1920-28. Returning to the Bureau of Public Roads, Edwards distinguished himself in his specialty. He retired in 1944 and for many years devoted himself to historical research in the history of bridges. After his death on May 10, 1952, his extensive historical files were donated to the Smithsonian Institution where they are still being used by scholars.

The Evolution of Early American Bridges*

BY

Llewellyn N. Edwards, C.E., D. of E.

(*Abstract of paper read at Caxton Hall, Westminster, March 15th, 1933*).

Development of Transportation.

" *Works great in one age of the profession will be superseded by the greater experience of future times.*" *John B. Jervis* (1869).

The earliest settlements in America and their communities were located upon sheltered bays, tidal inlets and navigable rivers, and streams providing communication by water between them; a condition of things tersely described in the adage—" The water was the first American highway." The expansion and multiplication of settlements created need for communication by land and primarily resulted in trails for pedestrian use only. Incidentally many Indian trails—narrow paths used by aborigine inhabitants—were utilized. We may, therefore, assume that the first bridges in America were of the foot-bridge type. Gradually changing conditions transformed the trails, by widening and straightening, into bridle paths, and horses became the common burden bearers with the saddle, pillion and packsaddle as equipment. In its turn this mode of transportation gave way to the two-wheeled cart adapted in its earliest stage to ox-team traction, while horses continued to be used for travel service of persons. The use of wheeled vehicles resulted, naturally, in the widening of the bridle paths and the removal of stumps, large stones and other obstructions from their surfaces.

The earliest foot-bridges consisted, rather commonly, of a single tree trunk felled at a convenient location across a stream, and the improved bridge of this type was produced by the addition of either one or two tree trunks laid parallel and close against each other. For greater convenience and safety the top portions of the timbers were sometimes hewn to provide flat surfaces and a " rayle " composed of a sapling made fast upon posts was placed along one side only. When the width of the stream to be crossed exceeded the length of available timbers, a multiple span structure was built, having one or more supports located within the stream prism, between the abutment supports. The

*Transactions of Newcomen Society, Vol. XIII, 1932-33

type, strength and rigidity of these supports or piers depended mainly upon the flood stage character of the stream and the desired service life of the structure. At times they were decidedly temporary and uncertain, while in other instances they were constructed with ample consideration for safety and permanence. For the latter, timber cribs founded upon the bed of the stream and ballasted with stones were common. To prevent undermining and displacement by stream scour these cribs or piers were frequently protected by stone riprap deposited upon the stream bed adjacent to their bases.

The horse-bridge differed in only two essentials from the foot-bridge. Its width was generally 5 to 7 ft., and the stringer timbers, instead of being laid in close contact, were separated and short lengths of small tree trunks hewn upon one side and notched to fit upon the stringers were laid transversely thereon to form a floor. A " rayle " was constructed on one side conforming with foot-bridge construction. (See Plate I).

The earliest cart-bridges were essentially a modification and a further development of the last-mentioned type. They were built to support increased load concentrations and commonly had " rayles " on both sides. The width between " rayles " was 10 to 12 ft., to allow sufficient space for a yoke of oxen with the teamster alongside. The gross weight of the load hauled was limited by the surface condition of the cart road, and, therefore, it is probable that about three tons was its maximum, with one to two tons as a usual limit.

In passing we may regard bridges for foot, horse and cart service as the beginnings—the pioneers—of American types of bridge construction. They were " a means to an end " in the crossing of relatively small rivers and streams, ferries being used for the larger crossings. The increased density of population attended by the developments of varied industries rendered necessary a corresponding remodeling and improvement of transportation facilities in which the two-wheeled vehicle took on another axle and the four-wheeled vehicle thus had its beginning in American highway service. The pioneer types continued to mitigate the difficulties of transport of the thinly settled " pioneer fringe " forging its way northward, westward and southward.

The versatility with manual operations engendered by the pioneering life of the period and the self-reliance, initiative and resourcefulness it fostered are evidenced by the development of types of structures wherein the art of carpentry was a major factor. Transportation developments were but a challenge to those engaged in bridge-building to produce structures which could be of service in spanning the major waterways. In a field so fertile and inviting for inventive ingenuity, it was natural that men of special genius and self-reliance competent to profit by their own mistakes, should direct their efforts to the development of types of structures involving the application of scientific considerations rather than the sort commonly described as " judgment founded upon error."

WOODEN BRIDGES.

Pioneer Bridges.

The records describing early bridges too commonly fail to give details of type and construction dimensions. However, we may glean from them valuable information.

The earliest authentic record of bridge building in America is contained in " The Generall Historie of Virginia, New-England and the Summer Isles . ." by " Captaine Iohn Smith, some tymes Governour in those Countryes & Admirall of New England," p. 195, which is here quoted:—

" 1620. With the first of the second yeere were held the Assises, where all the Bailiffes were fined for not giuing a beginning to the building of the bridges; there was also an order to restraine the excessiue wages all handicrafts men would haue: . . . The three bridges appointed by the general assembly, was followed with such diligence, though they were more then an hundred, or an hundred and twenty foot in length, hauing the foundation and arches in the Sea, were raised and accomplished, so that man or beast with facility might passe them." (See Fig. 24).

These bridges were built in the Summer Isles (i.e., Bermuda Islands) during " *The Gouernment of Captaine Nathaniel Butler.*"

In 1634 Israel Stoughton secured a grant from the General Court of Massachusetts Colony permitting the construction of a bridge over Neponset River at a location known as Neponset, where he had a mill. The nearest settlement was located four miles away. This bridge connected the towns of Dorchester and Milton. In May, 1654, Mrs. Stoughton (widow of Israel) petitioned to be relieved from maintaining a horse-bridge. The Court ordered that she maintain a " good foot-bridge with a good rayle." In 1655 the bridge was wrecked, probably by a freshet and was replaced by a cart-bridge.

In 1635 a foot-bridge was built at Ipswich, Massachusetts, probably over Town River. In 1641 it was rebuilt and widened and subsequently in 1663 and again in 1700 rebuilt and widened. In the last instance it is referred to as " Great Bridge."

In 1641 a foot-bridge was built over Charles River in Watertown, Massachusetts. " It is claimed that the first mill established in this country was located in Watertown " in close proximity to this bridge, " its dam forming the limit to salt water navigation from below." Here was constructed in 1648 the first horse-bridge built in Massachusetts Colony. This was called the "Mill Bridge," or "Great Bridge."[1]

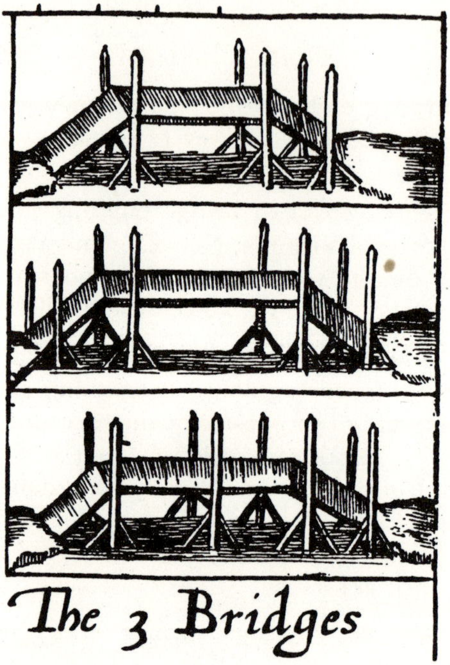

FIG. 24. WOODEN BRIDGES IN THE BERMUDAS, 1620.
From Capt. John Smith's *Generall Historie of Virginia* (Fifth Book), 1624.

In 1662 "Great Bridge" was built over Charles River between Cambridge and Brighton, Massachusetts.[1] This was a cart-bridge supported upon timber cribs. It was wrecked by a flood in 1685 and rebuilt in 1690. "For one hundred and thirty years this was the only bridge across the river for the accommodation of travel between Boston and the country lying to the north and west."

PILE AND FRAME TRESTLE BRIDGES.

Doubtless in many instances small wooden piles were used in the construction of pioneer bridges. The sapling size tree trunks were forced into the soils

[1] "An Historical Account of Some Bridges over the Charles River," by Lewis M. Hastings. *The Cambridge (Mass.) Historical Society Publication No. 7*, 1911.

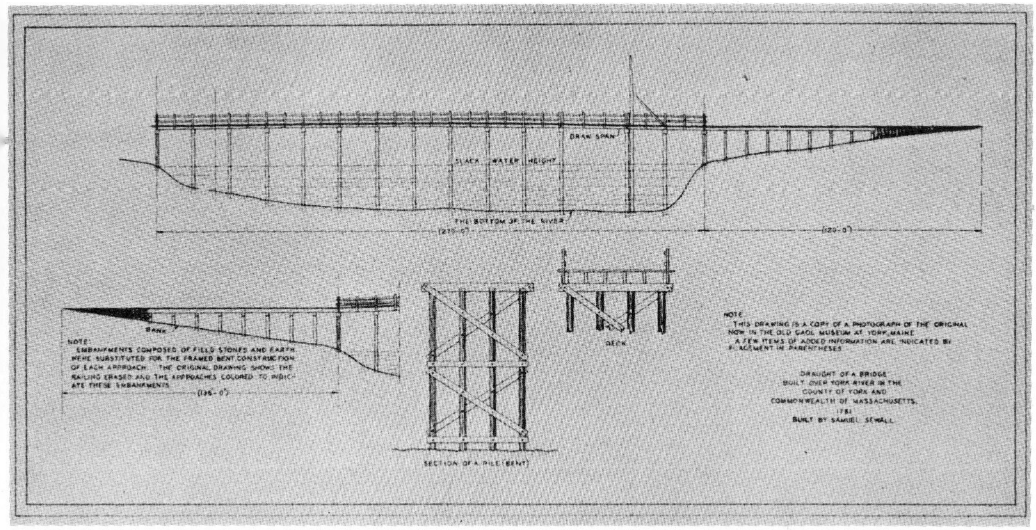

Sewall's Bridge, York River, Maine, 1761
From the original now in the Old Gaol Museum, York, Maine

Horse Bridge, New Hampshire
From a Pen and Ink Sketch based on Notes by the Author

Boston-Charlestown Bridge, Charles River, Massachusetts
From *Massachusetts Magazine*, Vol. I, 1789

Bellows Falls Bridge, Connecticut River, New Hampshire
From a Pen and Ink Sketch based on Material collected by the Author

of swamps, marshes and stream beds by using manually operated wooden mauls —a type of two-faced wooden hammer then in common use.

In 1761 Sewall's bridge was built over York River, York, Maine, by Major Samuel Sewall, a civil engineer of that place. This bridge is important in the annals of American bridge building, because it was the first constructed in accord with a plan predicted upon a survey of its site; furthermore, it was the first pile trestle structure. Its substructure consisted of 13 bents containing four piles each. It was "two hundred and seventy feet long by twenty-five feet wide, exclusive of the wharfs at each end, which reach to the channel." (See Plate I). "The piles or posts were of different lengths; the length being determined by probing the bottom of the river or mud with a pointed iron affixed to a long pole—a whole section containing four piles was framed, well braced, and the cap sill fastened on. At still tide it was floated to its place, and, by dint of labor, set upright and guyed. Large and heavy oak logs, the tops or lighter ends of which were secured inland, were then made use of, and the butts raised by tackles to a proper height; and by the striking of detents or latches, the ropes were released, the logs fell with great force upon the caps, and by their impetus this section was driven to the depth desired." A draw span "provided a sufficient way for sloops to pass and repass through the bridge."[2]

In 1785-6 the Boston-Charlestown, Massachusetts, bridge was built over Charles River (See Plate II). In the town records we find:

"April 27, 1720. Voted, that it be given as an instruction to the representatives of this town, now chosen to promote the building of a bridge over Charles River, at the place where the ferry hath been usually kept, viz., between Mr. Gee's and Mr. Hudson's point, and at no other place." That this project had been long contemplated is evident. An Act of Legislature March 9, 1785, provided for the formation of a company and "Board of Proprietors" to finance and manage the project.

The successful construction of the bridge over York River accounts, no doubt, for the selection of Major Samuel Sewall to design and superintend this important work, "which at the time was considered the greatest that had ever been projected in America." Lemuel Cox was "the master workman."

The following description indicates that apart from changes and additions rendered necessary by local conditions and increased width of structure, the general design was similar to that incorporated in Major Sewall's first pile bridge. That the same method of pre-assembling the pile bents (piers) and driving them as single units was used is evident.

"The abutment on the Charlestown side, from the landing, is 100 feet;

[2] *Ancient City of Georgeana and Modern Town of York*, by George Alex. Emery, 1873.

span to the first pier 16½ feet; 36 piers at equal distances to the draw 622½ feet; draw 30 feet; 39 piers on Boston side 672 feet; span to the abutment 16½ feet; from that to the landing 45½ feet; whole length 1,503 feet.

"The 75 piers total upon which this elegant structure stands, are each composed of seven sticks of oak timber, united by a cap piece, strong braces and girts, and afterwards driven into the bed of the river and firmly secured to a single pile on each side, driven obliquely to a solid bottom. The piers are connected to each other by large string pieces which are covered with 4-inch plank. The bridge is 42 feet in width and on each side is accommodated with a passage 6 feet wide railed in for the safety of people on foot. The bridge has a gradual rise from each end, so as to be two feet higher in the middle than at the extremities. Forty elegant lamps are erected at suitable distances, to illuminate it when necessary. There are four strong stone wharves connected with three piers each, built at the sides of the bridge. The floor of the bridge at the highest tides is 4 feet above the water, which generally rises about 12 to 14 feet. The distance where the longest pier is erected, from the floor of the bridge to the bed of the river, is 46¾ feet."[3]

In 1631 Harvard College established a ferry between Boston and Charlestown, the revenues from which were used "to defray the tuition of indigent students." In the charter provisions for the bridge this condition was carried forward by a requirement that the proprietors should pay annually $666.66 "in consideration of the income loss from ferriage."

The construction of the Boston-Charlestown bridge was followed by the construction of other improved structures of the same type, among which were the following:—

In 1788, Salem-Beverly, Massachusetts bridge; length, 1,484 ft.; width, 32 ft.

In 1790, Great Bridge, Cambridge (rebuilt).

In 1793, West Boston bridge; length, 3,484 ft.; width, 40 ft.

Doubtless the crudely improvised log pile drivers used by Major Sewall were entirely hand-operated, and that they were superseded by pile drivers involving the use of heavy wooden weights contrived to be lifted and let fall within a framework seems equally probable. In their earliest adaptations these machines may also have been entirely hand-operated. The horse-operated pile driver subsequently came into use marking a highly advantageous forward step in bridge construction.

In 1784-5 Bellows Falls Bridge over Connecticut River between Rockingham, Vermont, and Walpole, New Hampshire, was built by Colonel Enoch Hale. The natural ledge at the bridge site forms a deep gorge within which

[3] *Massachusetts Magazine*, September, 1789.

an " island rock " . . under which the highest floods pass without detriment . . " divides the stream into two channels, each about 90 feet wide." (See Plate II). The bridge was 368 ft. long and 60 ft. above high water mark. It appears to have been of framed trestle type with bents located upon the " island rock " and the ledge walls of the gorge from which inclined braces or struts were framed into the superstructure spans to give them additional rigidity and strength. To this end the superstructure was located at an elevation adequate to provide an effective sub-division of the span lengths, otherwise the bridge could have been of shorter length. This was the first bridge built over the Connecticut river, and was the only one until 1809, when the Toll Bridge, Springfield, Massachusetts, was built. It remained in service until 1840, when it was replaced by a covered wooden truss structure.

Arch and Truss Bridges.

The development of long span bridges involving the structural adaptation of arches and of trusses as the sustaining members widened the field of bridge construction and marked the beginning of a period of bridge building probably unprecedented in the history of any country. Bridges multiplied by the hundred; spanning waterways where crossings by fords and ferries had previously been the only means available. The formative processes which accompanied this development accomplished a departure from earlier methods of construction and evolved the beginning of bridge design and construction founded upon scientific engineering principles.

In 1737 Whiting's Bridge over Shetucket River, Norwich, Connecticut, was built by Capt. William Whiting. It was rebuilt in 1751, and was then 200 ft. in length. This structure was " swept away by a freshet in 1762." In 1764 it was rebuilt and renamed Leffingwell's Bridge. Its construction is described in the following quotation:—

" June 20, 1764, Leffingwell's Bridge over Shetuck River at Norwich Landing is completed. It is 124 ft. in length and 28 ft. above the water. Nothing is placed between the abutments, but the bridge is supported by Geometry work above and calculated to bear a weight of 500 tons. The work is by Mr. John Bliss, one of the most curious mechanics of the age."[4]

One may question the design load calculations, but the evidence is strong that some sort of truss structure was used for " nothing is placed between the abutments." Furthermore, diagonal truss members could form triangles and would suggest to the inexperienced, " Geometry work." In the absence of more complete and definite information, this bridge may be considered to be the first application of the truss to bridge work in America.

[4] *History of Norwich, Connecticut,* by Francis M. Caulkins, 1866, but there quoted from a newspaper account.

In 1792 Essex-Merrimac Bridge located near Newburyport, Massachusetts, over Merrimac River was built by Timothy Palmer. It consisted essentially of two bridges, the river being divided at the location by Deer Isle. While a considerable portion of the total length was composed of wooden cribs ballasted with stones and stringer type superstructure spans, there were two arched-truss spans located over the navigable channels: that on the Newbury side had a span length of 160 ft. with a height of 40 ft. above the high water elevation, while that on the Salisbury side had a span length of 113 ft., the underclearance being the same. The total length of the two structures, not including Deer Isle, was 1,030 ft. (See Plate III). In 1810 the 160-foot span was replaced by a chain suspension span 244 ft. in length center to center of towers.

In 1794 Piscataqua Bridge, located seven miles above Portsmouth, New Hampshire, over Piscataqua River, was built by Mr. Palmer, assisted by Zenas Whiting, of Norwich, Connecticut, as master workman. Its length was 2,362 ft. of which the major portion was pile bent and stringer construction. Over the navigable channel having water 46 ft. in depth there was a "stupendous arc," the chord of which was 244 ft. 6 in., with a versed sine of 27 ft. 4 in. This arched-truss span was 18 ft. 3 in. in depth and was constructed with three concentric ribs, the floor system being supported upon the middle one. Each rib was made up of five curved timbers composed of individual sticks hewn from crooked logs and framed to produce firm abutting joints, staggered to secure rigidity and strength.

Arched trusses having vertically curved floor systems were incorporated in the first structures planned by Mr. Palmer. The "over the hills" floor construction endangering life and property was modified to produce a slight rise in the roadway surfaces in his later bridges. The arched bottom ribs were made a part of all spans, but the top ribs were constructed horizontally in the later structures which in addition were provided with weather coverings to preserve their timbers from decay.

In 1804-5 Trenton Bridge, over Delaware River, Trenton, New Jersey, was planned and built by Theodore Burr. The sub-structure consisted of two abutments and four piers, all of stone. "The upstream ends of the piers were semicircular and after rising five feet gradually receded to the top where they were finished off in a half dome. These piers were 62 ft. long and 20 ft. deep. The bridge was 1,008 ft. long from the Jersey abutment to the Pennsylvania one which rested on Delaware Works Island. Its width was 36 ft. The superstructure . . . consisted of five wooden arches, respectively 203, 198, 161, 186 and 203 ft. in the clear, each composed of five great arched ribs rising from the chord in the proportion of 13 ft. to 100. These ribs were made of 4-inch pine planks, one foot wide and from 35 to 50 feet long, built up into a thick laminated rib 3 feet wide. The relative placing of these ribs left two openings of 11 feet each in the center of the bridge for carriage ways and two more, each 4 ft. 6 in. wide on the sides for foot walks. The ribs were spaced and bound together on the top circumferences of the arches by ties and diagonal braces fastened to the ribs by bolts and screws at intervals of eight feet. The floor was suspended from the ties by perpendicular iron rods securely fastened in the wood. Wing arches and diagonal braces were efficiently used throughout to eliminate all motion between the parts of the bridge."

The entire bridge was covered, and the covering of its ends formed "arched doorways over the carriageways and foot-walks." Balustrades 4 ft. high for the protection of pedestrians were placed upon the outer edges of the foot-walks.

In 1814-15 McCall's Ferry Bridge, McCall's Ferry, Pennsylvania, was built over Susquehanna River by Theodore Burr. No record exists of the total length of this structure. However, its most interesting portion was the wooden arch span, the dimensions and method of erection of which are preserved in a letter written by Mr. Burr, quoted here in part:—[5]

"This arch is without doubt the greatest in the world. Its length, between the abutment and the pier, is three hundred and sixty feet four inches, the chord line of the arch, three hundred and sixty seven feet. The width of the main part of the bridge is thirty two feet; the wings of the pier spread eleven feet eight inches on each side, which makes a base of fifty five feet four inches. At the abutment the wings spread seventeen feet each, which makes a base of sixty six feet. The altitude or rise of the arch is thirty one feet. The arch is double, and the two segments are combined by King posts seven feet in length between the shoulders, and are united to the arch by lock-work. Between the King posts are truss braces and counteracting braces. . . . The center of the arch is sixty one feet from common low water to the lower and seventy feet four inches to the upper segment." Concerning the river at the site, "it is only six hundred and nine feet in high water; and in low water the whole river runs in the space of three hundred and forty eight feet. In this state it has been sounded by Doctors Preston, Marshall and Baley, gentlemen interested in the bridge, and ascertained to be one hundred and fifty feet in depth."

The original intention was to build the arch complete upon eight floats "lying in the water, ranged along the shore nearly a quarter of a mile below the abutment. The floats were placed at proper distances with their ends to the shore, and on each there were raised two bents or frames, varying in height to correspond with the curve of the arch" thus supporting the structure at an

[5] From Mr. Burr's letter, dated Harrisburg, Pa., Feb. 26, 1815, and printed in *Niles Weekly Register,* Vol. 9, 1815.

elevation of approximately sixty feet above the river surface to float it into its position upon its substructure. The construction of the arch span was begun on October 1, 1814, and "On the 17th of December we had the whole in readiness to move up to the abutment, and on the same day the anchor ice began to run a little."

Alternating cold and mild weather in the latter portion of December produced in the gorge an accumulation of sheet and anchor ice which jammed against and under the floats and ultimately "hove in a mountain of ice upon us, the average height of which for about a mile above and below us was ten feet above the surface of the water at the shores. . . . At the same time the whole body of ice moved down from twenty-five to thirty feet which bore so hard against the floats that it broke and smashed more than half of them. . . ." Ultimately a condition developed considered to be favorable for moving the arch upon runners over the top of the ice pack and leveling its rough jagged surface was at once commenced. The arch was cut at its center and by the use of eight capstans and man power in plenty the halves of the span were moved upon false-work over the ice at an elevation of over 60 ft., adjusted in position and united on January 30, 1815. Not only is this the earliest record of an attempt to transport an arch span upon floats or scows, but in addition is unique in being the first and only arch span hauled into position upon ice.

In 1812 Upper Ferry Bridge over Schuylkill River, Philadelphia, Pennsylvania, was built by Louis Wernwag. The dimensions of the massive masonry abutments were:—Easterly: 60 ft. on front; 40 ft. thick; 48 ft. high; founded on rock. Westerly: 62 ft. in front; 40 ft. thick; 40 ft. high; founded on 599 piles, platformed to provide a masonry foundation. The superstructure was "A single arch whose chord is three hundred and forty feet three inches and three quarters of an inch. Versed sine, or rise from the chord to the apex nineteen feet eleven inches. The curve formd by this arch is not, however, that of the segment of one great circle; but of segments of circles of unequal radii. Those of the longest radii being nearest the abutments and the shortest at the apex."

"The width of the superstructure at the abutments is fifty feet and at the apex thirty-two feet. The three ribs extending across the river from one abutment to the other are set in cast iron head blocks, upon an offset in the wall, twelve feet above common high water mark. Their distance asunder, at the abutments is twenty-one feet and at the apex thirteen feet one inch. Each of these ribs is composed of six small ones, in thickness six inches and of the average depth of fourteen inches. The small ribs are placed on their edges, two in breadth and three in depth, and so formed as to be at the abutments equal to a solid mass of timber four feet deep and one foot thick, gradually diminishing in size so as to be at the apex but three feet deep and one foot thick. They are prevented from coming into contact by one inch iron bars placed between them, six feet asunder, but are connected together by large iron bands also six feet apart. . . ."

To maintain the ribs " in their proper relative situations " adequate cross ties and braces were provided " so as to completely prevent any lateral motion." " Upon each of the ribs are twenty nine King posts, set in cast iron boxes of a proper shape to receive them, and secured by iron stirrups embracing the rib and post. Upon the tops of these posts are truss ties extending the whole length of the bridge, and upon the truss are ties extending across the bridge, well secured to the King posts and truss ties.

" The King posts, truss ties and cross ties are secured in their proper places by iron bars extending through the mass of masonry in the abutments to the top, and from thence through the timbers longitudinally and diagonally from one end of the bridge to the other.

" The cartway at the entrance of the bridge is 14 feet above the spring of the arch, and passes between the ribs at its apex, forming in its progress an angle with the horizontal of about two degrees." (See Plate III). This bridge was burned, 1838.

The types of trusses incorporated in the Piscataqua, Trenton and other individual structures are shown in Fig. 25, which also shows other types to be described later. The arrangement of truss member in the latter is that indicated in patent drawings.

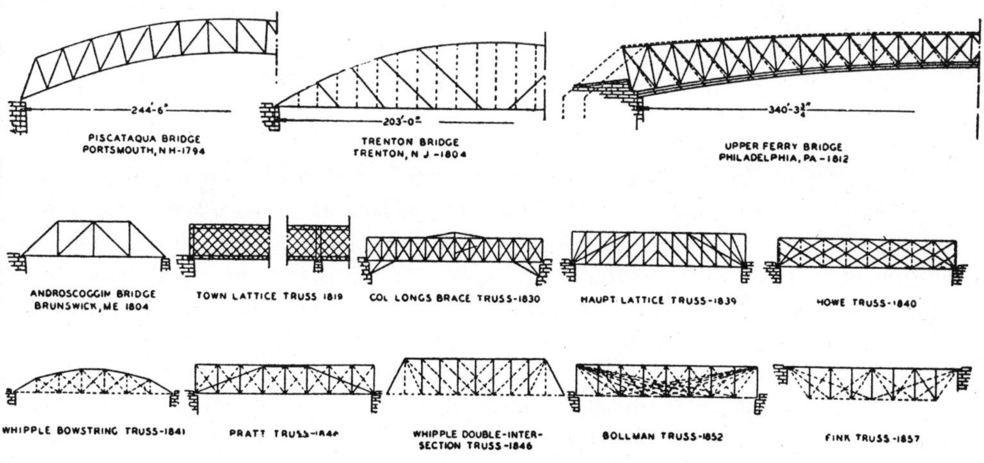

EVOLUTION OF AMERICAN BRIDGE TRUSSES

FIG. 25. TYPICAL AMERICAN BRIDGE TRUSSES.

In 1819 Ithiel Town developed a type of truss (patented Jan. 2, 1820), which departed rather widely from common practice and blazed the way for future development. It consisted essentially of top and bottom chords placed parallel, with vertical posts located at the truss ends and with inclined web members forming a lattice work. The timbers throughout were of the dimensions commonly termed plank, in lumber, and were, therefore, readily obtainable. The chords consisted of two or more sticks laid parallel and upon opposite sides of the latticed web. In long spans each chord consisted of two elements placed one above the other. To secure an effective transfer of the truss loads to the abutments the trusses were extended practically their depth distance beyond the breast walls and vertical posts were placed at the extreme truss ends and near the fronts of the abutments. The trusses were built continuous over piers and vertical posts were located above the pier bearings. The spacing of web members was varied with the lengths of span as was also the depth of truss and the sizes of timbers. The timbers (planks) were commonly 2 to 4 in. in thickness and 10 to 14 in. in width. All web member intersections and their junctions with the chords and posts were held firmly by wooden trenails, usually two in web intersections and three in chord and post junctions. The floor system was supported upon the truss chords and the trusses were secured laterally by struts and diagonal bracing.

The framing and assembling of these bridges involved only the simplest of mechanical operations and was executed without the use of bolts, rods, or straps of iron. The entire structure was preserved from decay by being roofed and weatherboarded upon the sides and gables. Mr. Town built his first bridge "about two miles north of the center of New Haven (Conn.) over the Mill River."

It is very probable that more bridges were built of the Town type than of any other wooden truss. Very few of these had span lengths in excess of 200 ft.

In 1830 Colonel Stephen H. Long developed and patented a parallel chord truss bridge which he improved in 1836 " to produce and insure lateral stiffness and inflexibility in wooden or frame bridges." Of the original type Jackson Bridge, built 1830, over the Baltimore and Ohio Railroad on the Washington (D.C.)-Baltimore (Md.) road was the first. Its length was 109 ft., width 24 ft., and height of trusses 14 ft. Many bridges of the Long type were subsequently constructed.

In 1840 a truss bridge of the parallel chord type was patented by William Howe. The arrangement of the truss members permitted a complete stress analysis by the mathematical practice then in use. All previously patented truss devices were inadaptable to analysis, indicating rather clearly the absence of mathematical considerations other than, perhaps, those of an empirical kind. In 1844 Thomas W. Pratt and Caleb Pratt patented a parallel chord and also a cambered top chord type. In 1846 Mr. Howe " devised an improved truss bridge in which the arch beam is combined with the general truss in such manner that it may be made to co-operate with the other parts of the truss frame. . . ."

All of these types were favorably received in construction practice.

Following the adoption of cast and wrought iron as structural materials a comparatively few bridges were built in which these materials were used in combination with wood. Wrought iron for web members subjected to tensile stresses had been in use for several years. However, that structures should appear during the transition period in which wrought iron was used for bottom chords and other important tension members with cast iron for joint casting and " all or most of the compression members of wood " was a normal condition of development. Seven patents were issued by the United States Patent Office for such " combination " bridges.

METAL BRIDGES.

The credit for inventing the first American metal bridge justly belongs to Thomas Paine, who when residing in Philadelphia in 1787 " proposed to erect a bridge over the Schuylkill River, and that it should be a great span, without piers (so as not to be obstructed by ice). Paine boldly offered to build an iron bridge of 400 feet span." Letters patent for his invention were granted by King George III, August 26, 1788. In that same year he arranged with Messrs. Walker, of Rotherham, Yorkshire, to have the design of his bridge carried out. Following the success of a test made upon two arch ribs of 90 ft. span and 5 ft. rise, Mr. Paine had a complete bridge involving five arch ribs of 110 ft. span and 5 ft. rise, fabricated and sent to London, where it was exhibited for some time in a bowling green at Paddington " as a specimen for establishing a manufactory of bridges to be sent to any part of the world." Soon afterward his activity in the French Revolution superseded that in regard to bridges. Financial difficulties rendered him unable to defray the expense and the manufacturers utilized his bridge in the construction of one in the British Isles.

Wrought Iron Suspension Bridges.

In 1801 a wrought iron suspension bridge was built over Jacobs Creek near Greensburg, Pennsylvania, by James Finley. " Its construction is on principles entirely new. . . . It is solely supported by two iron chains extending over four piers 14 feet higher than the bridge fastened in the ground at the ends, describing a curve line touching the level of the bridge in the center. The first tier of joists are hung to the chains by iron pendants or stirrups of different lengths so as to form a level of the whole. The bridge is 70 feet span and 13 feet wide; the chains are of an inch square bar in links from five to ten feet long; but so that there is a joint where each pendant must bear." The suspenders constructed to fit upon the link ends were held in position by cleverly contrived clamps and their bottom ends were stirrup shaped to engage the "needle beams of the platform" or floor system, the center panel beams of which were supported directly upon the ends of the center links. (See Plate IV). Patent rights were granted to Mr. Finley June 17, 1803. Prior to 1811 eight bridges of his design had been constructed.[6] It is generally conceded that James Finley originated the suspension bridge as it is now constructed, that is to say, with a horizontal floor system suspended from suspension members. However, Capt. Basil Hall, F.R.S., describes in his *Journal of Travels in South America*, 1820, bridges of this type built by the aborigine inhabitants wherein the cables and suspenders were ropes made of "the raw hide of bullocks." (See Plate IV).

In 1810 the suspension span of the Essex-Merrimac bridge, replacing one of its wooden arched-truss span (see page 102) was built by John Templeman, a lessee of the patent rights of James Finley. The span was 244 ft., center to center of towers; the towers 47 ft. long and 37 ft. high, of wooden construction battered upon their sides to conform to the batter of the piers and protected by weatherboarding. Ten suspension chains, each 576 ft. long, arranged to provide three in each outside member and four in the center member supported the floor system, 30 ft. in width and divided into two roadways by the center suspension member. "The chains generally are formed of links about 2 feet long, made of 1-inch square iron."

In 1811 Horn's Bridge over Lehigh River, Easton, Pennsylvania, was built of the Finley type. "This bridge was suspended on four chains, hanging in two loops and two half loops, having two passways for teams and a footwalk between, which was guarded by hand railings." It "is the first suspension bridge on record having more than one span." This bridge replaced a wooden arch bridge of 280 ft. span, built by Abraham Horn, 1797.

The establishment of wire manufacturing plants in Connecticut, Massachusetts and Pennsylvania, resulted in the gradual substitution of that material for bar iron in suspension bridges. Wire was first used in the suspension members of a foot bridge over the Schuylkill River at Philadelphia, January, 1816. This bridge had a length of 408 ft. between supports and its footway was 18 in. wide. It was intended only as a temporary structure, but was in service nearly a year.

[6] *A description of the Chain Bridge,* by Judge James Finley, of Fayette County, Pa. Pamphlet dated April, 1811.

In 1842 Upper Ferry Bridge, Philadelphia, Pennsylvania, was replaced by a suspension bridge designed by Colonel Charles Ellet, who was the first American engineer to adopt wire in constructing highway suspension bridge cables. However, no evidence exists that he claimed originality therein, but instead his pamphlet *A Popular Notice of Suspension Bridges, etc.*, Philadelphia, 1843, gives to M. Seguin, "a practical mechanic" of Lyons, France, full credit for this invention in 1823. Construction was commenced in July, 1841, and the bridge opened for traffic January 1, 1842. Its length between points of suspension was 357 ft.; its width 26 ft., providing a roadway 18 ft. wide and two sidewalks each 4 ft. wide; its length of roadway proper was 342 ft. "The flooring is upheld by ten cables of iron wire resting on cast iron rollers placed on the summits of these columns which move sufficiently to compensate for any change—consequent on changes of temperature or other causes and relieve the columns of all injurious strain." The columns were of granite masonry 8½ ft. square and 30 ft. high.

"The process adopted in the manufacture of the cable was such as to secure beyond all doubt, an equality of tension among the separate strands. The cables are formed of an assemblage of strands of wire about the diameter of a common writing quill (No. 10 in commerce) which are laid parallel to each other . . . and afterwards collected in a solid mass and bound together by ligatures of the same material. A coat of varnish is applied to every wire previously to the formation of the cable which offers . . . an adequate protection against oxidation."

The ten cables contained 2,816 strands of No. 10 wire. "In lieu of a brace to guard against horizontal vibrations the cables are inclined toward the axis of the bridge . . ." It is probable that this is the first application of cradled cables in American practice. The roadway system was reinforced by longitudinal timbers to secure rigidity against vertical deformation. It was "suspended from the cables by means of wire cords similar in their construction to the suspension cables, though of much smaller diameter, distributed along the curves at distances a fraction over four feet from each other, and corresponding to the joists of the flooring."

In 1847-49 Col. Ellet designed and superintended the construction of the Wheeling and Belmont Bridge Company's bridge over Ohio River at Wheeling, West Virginia. The construction is similar to that already described. However, in this instance the span length is 1,010 ft. between the points of suspension and provides an underclearance of a fraction over 93 ft. "immediately over the river channel" commonly used by navigation.

The next longest suspension bridge built was the international structure over Niagara River connecting Lewiston, New York, with Queenstown, Ontario.

Its length is 1,042 ft. between points of suspension. It was designed by Edmond W. Terrell and built in 1852.

Both of these bridges remain in service (1933).

CAST IRON ARCH AND ARCH-TRUSS BRIDGES.

The arch in stone and in wood had demonstrated its service reliability and been accepted as a fundamental element in construction long prior to the early developments of bridge construction in the American Colonies. It was, therefore, made a part of many of the earliest bridges involving long spans and its adoption in metal bridge design was essentially a normal procedure, notwithstanding the condition that parallel chord wooden trusses were then accepted as good practice.

In 1836-7 Dunlap's Creek Bridge between Brownsville and Bridgeport, Pennsylvania, was designed and constructed by Captain Richard Delafield, Corps of Engineers, U.S. Army, assisted by Lieuts. George W. Cass and William R. McKee. It is composed entirely of cast iron, except the railings, which are wrought iron. In a report to Brig. Gen. Charles Gratiot, Chief of Engineers, Capt. Delafield described the design as "differing in its principles of construction from any of which I could find a notice by either English or French Engineers." The span of the arch is 80 ft.; its rise 8 ft. It consists of five ribs of elliptical cross section, spaced 5.77 ft. apart and having "spandrel supporters" resting thereon which are the direct supporting members of the flooring plates. The roadway width is 24.75 ft. The construction involved masonry 4.882 perches; cast iron 322190 lb.; wrought iron 4,714 lb.[7] This bridge is still in service.

In 1840 Squire Whipple built his first iron bridge over the Erie Canal. It was of the bow-string truss type in which the arch was composed of chordal cast iron segments constructed with projecting lugs "to assist in keeping the ends in place." In cross section these segments consisted of two right angle shaped portions rigidly connected by integrally cast (T-shaped) struts or cross bars located 3 to 4 ft. apart. The transverse width of each segment varied, the arch being widest at its "heel" ends and tapered to the center of its length, thus obtaining transverse rigidity. The web members and bottom chords were of wrought iron; the latter consisting of a chain having link lengths equal to the truss panel lengths and engaging the trunnions forming a part of the joint castings. Subsequently a large number of bridges were built of this type having spans 50 to 180 ft. in length. It was patented 1841.

CAST IRON TRUSS BRIDGES.

The successful results attained by the use of wooden parallel chord truss bridges led to the development of similar structures wherein all tension members were of wrought iron and all compression members of cast iron.

[7] Official records of the Corps of Engineers, U.S. Army.

In 1833 August Canfield patented a parallel chord metal truss bridge. Its uniqueness consisted of anchoring its top chords into the abutments and arranging its web members to produce tension in both the upper and lower chords. No record exists that a bridge of this type was built.

In 1841 Earl Trumbull invented a parallel chord truss bridge, one of which was built in that year over the Erie Canal at Frankfort, New York. This was probably the first American metal parallel chord truss bridge. "The truss was a combination of the truss and suspension principles, and was formed of: First, seven cast iron sections or panels of about 11 ft. in length and 7 ft. in depth, cast solid, each segment consisting of an upper chord, a pair of diagonal braces and half of a hollow cylindrical post at each end (of the segment), except that the end segments had full cylindrical posts at the abutments. These semi-cylinders being bolted or clamped together in series, formed full cylindrical posts, which were flanged at the bottom, and through which were passed vertical bolts securing them to wooden transverse floor beams. Second, two wrought iron suspension rods ($1\frac{1}{2}$ in. diam.) attached to the top ends of the end posts, and sagged in a parabolic curve, so as to pass under and support the two centremost floor beams, and under lugs cast at the proper elevations upon the posts intermediate between the centremost and the end posts, whereby such intermediates were supported.

"Cross sections of the chords and diagonal braces (were) of the + formed section." The span length of this bridge was 80 ft.[8]

In 1846 Squire Whipple developed a trapezoidal truss bridge having a double intersection web system and inclined end posts—the first American type having these features and the starting point of their general use. The top chords, end posts and other compression members were of cast iron; while all bottom chords, web diagonals and other tension members were wrought iron.

In 1852 the Bollman truss was patented by ~~August~~ Bollman, and in 1857 Albert Fink patented the Fink truss. Both types are formed by superposing a number of inverted king post trusses. In the former type the superposed elementary trusses are of uniform depth and span, while in the latter they are of varying span and may be of varying depth. In neither type is there a true bottom or tension chord and this member was frequently omitted. Both types were used for through and deck bridges. However, although equally simple in construction the Fink type proved of greater merit in service.

The Howe, Pratt, Whipple, Bollman and Fink trusses were readily adaptable to the system of analysis promulgated by Squire Whipple, and in addition possessed merits which insured their wide adoption for metal structures.

[8] Quotation in paper *American Railroad Bridges,* by Theodore Cooper. *Transactions of the American Soc. Civil Eng.,* Vol. XXI, 1889.

Within the fifteen years preceding the Civil War, metal bridge construction made very creditable progress, but became practically dormant throughout that struggle (1861-65). It gained remarkable impetus within the two succeeding decades.

General.

The United States Patent Office issued from 1797 to 1860 inclusive fifty-one patents for differing types of wooden bridges, and from 1803 to 1860 inclusive sixteen for types of metal bridges of which only one, that for James Finley's chain suspension bridge, was issued prior to 1833. From these the test of time and the ever-changing demands of service needs have selected, by process of elimination, the most efficient and practicable, and thereby rendered obsolete and of interest only as matter of evolution and history a very large number.

That some of the patentees possessed scanty knowledge of the service functions of bridges is clearly indicated by the devices produced; some were cumbersome, massive and inadaptable, while others were replete with artistic curves and uncertain intricate combinations of wood or metal. Speaking generally, the types adapted to mathematical stress determination and containing no redundant elements were at the end of the period here considered accepted as good practice.

The truly remarkable development which marked the period 1797 to 1812 inclusive, gives evidence of a condition wherein men possessing sound judgment and resourcefulness directed their efforts to produce structures which would not only fulfil all requirements, but at the same time be superior to and more bold than any before. Without question or doubt these early bridge builders were mainly guided and encouraged by the knowledge of the results of experience, rather than by considerations of physical laws. However, the record of the immediately succeeding decade and more, when viewed backward through more than a century, leads one to wonder whether further achievement in the art of bridge construction was then dependent upon the application of mathematics. Whatever may have been the underlying factor, it is pertinent to note that within the period 1797 to 1812 inclusive ten patents were issued, and within the period 1813 to 1828 inclusive five were issued, but of the latter only one possessed real constructive and evolutionary merit.

An incident is related of the "broad-axe and chisel" period of bridge design wherein a contractor was about to intrust work to one of his foremen, but entertaining doubts as to his ability to determine the sizes of timbers and execute the framing, inquired whether he felt equal to the task. He was assured of that worthy's ability when the latter replied, "I have two infallible rules by which to decide such details. I first take the average judgment then guess and allow."

While we may entertain a belief that Timothy Palmer, Theodore Burr, Lewis Wernwag and other designers and builders of early wooden arch and truss bridges made use of mathematical rules for determining the structural elements of their bridges, no substantiating records exist in proof of this, nor have we positive evidence of this kind prior to about 1840.

In 1842 Herman Haupt, C.E., produced a small pamphlet, *Hints on Bridge Construction by an Engineer.* In 1847 Squire Whipple, C.E., produced *A Work on Bridge Building* consisting of two essays. The one elementary and general, the other giving *Original Plans and Practical Details for Iron and Wooden Bridges.* In 1869 Mr. Whipple produced a work *Bridge Building, being the Author's Original Work published in* 1847 *with an Appendix containing Corrections, Additions and Explanations, Suggested by Subsequent Experience.* In this book he states: "The original publication was a pioneer effort, . . . It is believed that no previous attempt had been successfully made to reduce Truss-Bridge construction to its simplest elements, and to determine by exact calculations, the forces acting upon the various parts of such structures and to deduce thence the proper sizes and proportions of such parts, upon known and reliable principles."

In 1851 Herman Haupt produced his work *General Theory of Bridge Construction containing demonstrations of the principles of the art and their application to practice; furnishing the means of calculating the strains upon the chords, ties, braces, counterbraces, and other parts of a bridge or frame of any description.* His autobiography *Reminiscences of General Herman Haupt* contains the statement that the manuscript of his work was completed five years prior to its publication.

The prefaces of these two books give credence to the belief that neither Mr. Whipple nor Mr. Haupt had knowledge of the other's efforts to reduce the theory of bridge design to the printed page. The following is quoted from Mr. Haupt:

"From the great importance of the art of bridge construction it might be supposed that its principles would be familiarly understood by all whose occupations require an acquaintance with it. If, however, any work exists, containing an exposition sufficient to account generally for the various phenomena observed in the mutual action of the parts of trussed combinations of wood or metals, the author has neither seen or heard of it."

This review of bridge construction and development directs attention to the composition of structures regarded as typical of their individual times and as, in general, marking stages of progress. The record is incomplete, for to have included all early bridges possessing real interest and value, each contributing to progress, and to have made mention of the early bridge builders who gained inspiration from emergencies and careful observation, giving to each his due credit, would have proved too extensive a task. However, the review may be considered as embracing in brief form, more than two centuries extending from the earliest settlement in the American Colonies to the sixties of the nineteenth century.

In retrospect we may consider that the early American settlements or colonizations came into being under political influences and circumstances considerably variant; but withal it may be said that whatever the original manner of territorial settlement, growth or advancement, the Revolution served as perhaps the most powerful factor in homogenizing of political thought and industrial activity and progress.

The paper was illustrated by thirty-two lantern slides from the author's collection of engravings, drawings and photographs, which served to amplify the descriptions in the paper. These illustrations are reproduced in part in Plates I to IV, and Figs. 24 and 25.

Discussion.

The President (Mr. H. W. Dickinson) said it would be noticed that the practice with these very early bridges in America was the same as that in Europe, including roofing over to protect from the weather, as is still to be seen in Switzerland. It would be desirable to know more about Paine's cast iron bridge of 1788, for it was an early use of that material. The interest of the paper was centred mainly on the truss bridge, with the development of which American engineers are so largely to be credited. Bridge development had been studied by another member of the Society, Professor R. S. Kirby, with P. G. Laurson, *Early years of modern civil engineering*, 1932.

Mr. Thomas Clarkson recalled that thirty years ago in the States he had driven over many wooden bridges. Three years ago he had stayed in Maryland, just over the border from Delaware, and across Bohemia Creek. There was a timber bridge in sight, and in hearing, too, for the bumping of cars over the floor planks disturbed him in the night; last year he found a concrete bridge had replaced it. The author had barely mentioned stone bridges and had not given any examples.

Eng. Capt. E. C. Smith drew a parallel between bridge building and ship building in wood. Wooden bridges up to 360 ft. span were mentioned and it was interesting to observe that America had built the biggest wooden ships. He cited the "Adriatic," 1857, the Collins Atlantic liner, 355 ft. long, 3,670 gross tonnage; the "Great Republic," *c.* 1859 for the Pacific service, 360 ft. long, 4,300 tons; and the "Dunderberg," 1865, a sea-going ironclad, 377 ft. overall,

ESSEX-MERRIMAC BRIDGE, MERRIMAC RIVER, MASSACHUSETTS, 1792
From *Massachusetts Magazine*, Vol. XXI, 1792

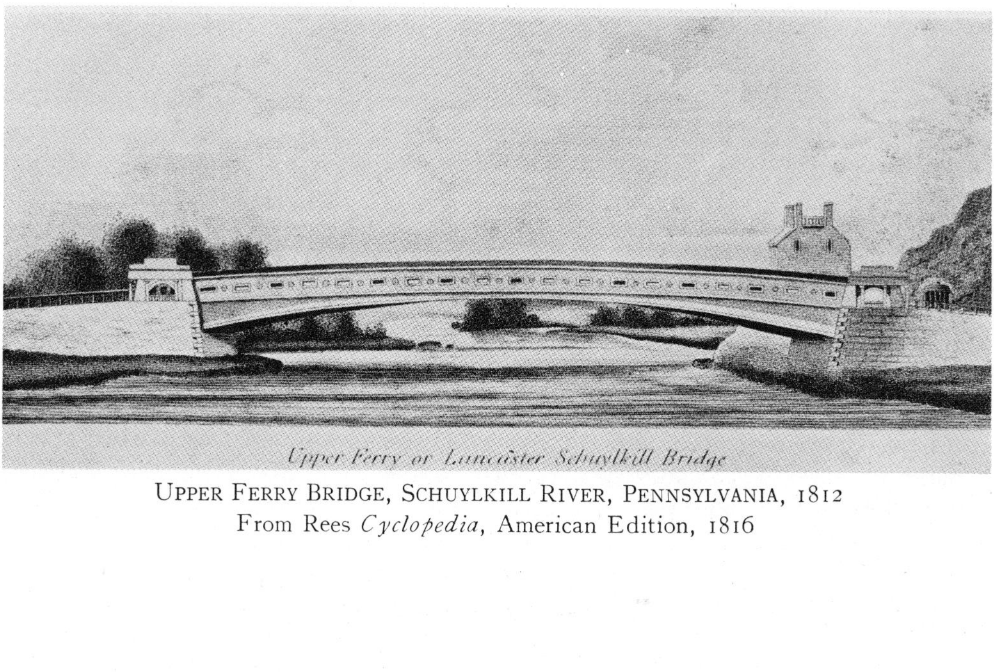

UPPER FERRY BRIDGE, SCHUYLKILL RIVER, PENNSYLVANIA, 1812
From Rees *Cyclopedia*, American Edition, 1816

ABORIGINES' SUSPENSION BRIDGE, CHILE
From Capt. Basil Hall's *Journal of Travels in South America*, 1820

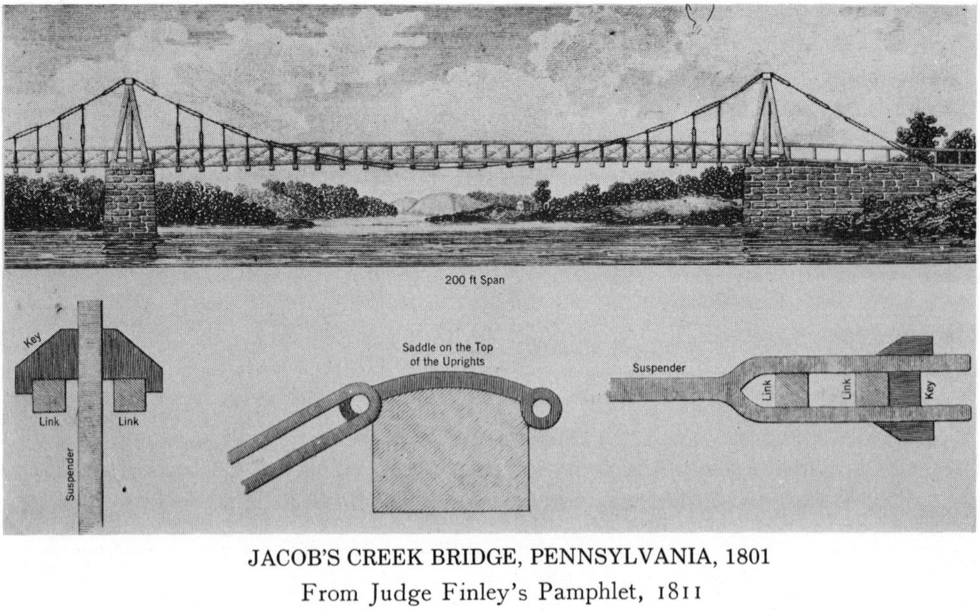

JACOB'S CREEK BRIDGE, PENNSYLVANIA, 1801
From Judge Finley's Pamphlet, 1811

7,060 tons displacement. Whether a wooden ship 500 ft. long could be built he did not know, but it seemed as if the limit of wooden ships had been reached at the same time as that of wooden bridges.

MR. S. B. HAMILTON commented on the sequence disclosed between technology and science. By the time the first treatise on truss bridge construction had been brought out, it appeared that every conceivable type had been tried and erected. What science did was to hold a post-mortem on the elucidation of the problems involved.

MR. A. TITLEY enquired the date of the hand-operated pile-driver described, and suggested that it had been taken over to America by the Dutch in the 17th century.

MR. REX WAILES drew attention to Howland's *Annals of North America*, 1877, in which would be found several references to bridges.

MR. LANCASTER BURNE said the wooden bridges put up by Brunel in Cornwall should not be overlooked.[9]

DR. A. P. THURSTON did not consider that it was the strength of timber, which weight for weight was as strong as steel, that limited the size of structures made of it, but the convenience of metal and the liability of timber to decay. For aircraft, for instance, there was much to be said for wood; silver spruce had a tensile strength up to 3,000 lb. per sq. in. and a compressive strength of 8,500 lb. per sq. in.

THE HON. SECRETARY (MR. A. STOWERS) disclosed that the material that Capt. Edwards had put before the meeting was only a fraction of what he had gathered. Mr. Stowers showed slides of the Iron Bridge near Coalbrookdale and several other bridges; in the case of a wooden bridge in Ceylon any member of the truss was separately removable.

CAPT. EDWARDS, in reply, drew attention to the condition that marked the early development of long span wooden truss bridges wherein builders appeared to pattern after and rival each other in attempts to outdo the accomplishments of earlier efforts.

An explanation of the clattering noise resulting from movements of the floor planks under traffic in wooden bridges is found in the fact that metal spikes were expensive and, in consequence, the planks were laid without fastenings upon the stringers or joists. This practice became common.

In reply to a comment concerning the maximum length of span that could be built of wood, it seems appropriate to mention that Thomas Pope, landscape architect, in his book, *A Treatise on Bridge Architecture*, 1811, proposed to build one of his "flying pendant lever" bridges as a single span structure over the Hudson River at New York. In view of the fact that the Brooklyn

[9] *Proc. Inst. Mech. Eng.*, 1899, p. 344.

Suspension Bridge, completed in 1883, has a span of 1596 ft. between supports, it is extremely doubtful whether Mr. Pope's bridge could have been built without collapsing during construction. Mr. Pope's book consisted mainly of a history of bridges and was the first book on bridges printed in America.

The kind of wood used for bridge construction, depended upon location: pine, spruce and oak were mainly used in the northerly States, while in the southerly States the hard pines, chestnut and oak indigenous thereto were used. With the development of wooden bridges, added efficiency was obtained by the provision of details rendering it possible to remove and renew decayed or otherwise injured timbers while the structure remained in service.

Stone masonry structures of moderate size were built in considerable numbers at comparatively early dates. Brick has been used in bridge work only to a limited extent, confined almost entirely to structures having very short spans.

Correspondence.

Mr. W. H. Thorpe wrote: Captain Edwards's interesting paper raises once again the question of the methods—if any—by which bridge-builders up to the end of the eighteenth century determined the scantlings of timber structures. These builders were clearly men of superior intelligence, and maybe not altogether lacking in some knowledge of strength estimation by the use of models. A scale model tested to destruction will, it is true, tell no more than the probable manner of failure, and gives, by this alone, no indication of strength margins of the full-size span, but when it is considered that for similar structures the total weight varies as the cube of the span, and the sectional area of its members as the square of the span, it becomes apparent that unit stress in the members due to the weight carried varies as the span simply. We have thus a means by direct test of models of forming an idea of the margin of strength, if any, of the full-size structure. By some such method of reasoning it would be possible to determine suitable scantlings with but slight knowledge of stresses as such.

BIBLIOGRAPHY OF BRIDGE BOOKS

Many books have been written on the history and development of bridges. The following bibliography includes books devoted to the general subject of bridge history, as well as a number restricted to covered bridges. While a few of the books are somewhat technical in nature, each of them contains information of interest to the layman.

There are many pamphlets published by various local Historical Societies and Associations of Commerce which give information about covered bridges in specific localities, such as those in Bucks County, Pennsylvania and Parke County, Indiana, both of which are nationally known for their covered bridges. A number of the maps published by State Highway Departments show the location of the State's covered bridges. In addition to the "World Guide to Covered Bridges" mentioned in this bibliography, the National Society for the Preservation of Covered Bridges also publishes a quarterly, "Covered Bridge Topics". Their address is given in the Introduction.

Allen, Richard Sanders, "Covered Bridges of the Northeast," Stephen Greene Press, Brattleboro, Vt., 1957, 121 pp. illustrated.

Allen, Richard Sanders, "Covered Bridges of the Middle Atlantic States," Stephen Greene Press, Brattleboro, Vt., 1959, 120 pp. illustrated.

Arcambeau, Edme, "The Book of Bridges," Gowans and Gray, London, 1911.

Barman, C.A., "The Bridge, "London and New York, 1926, 272 pp., illustrated.

Beckett, Derrick, "Bridges," — one of the books from the "Great Buildings of the World" series, Hamlym Publishing, London, 1969, 191 pp.

Bell, William E., "The Art and Science of Carpentry Made Easy," Philadelphia, 1857, 134 pp.

Billings, Henry, "Bridges," New York, 1956, 159 pp., illustrated.

Black, Archibald, "The Story of Bridges," McGraw-Hill Book Co., New York, 1936, 226 pp., illustrated.

Brangwyn, Frank and Sparrow, Walter S., "A Book of Bridges," John Lain Co., London and New York, 1914, 415 pp.

Congdon, Herbert Wheaton, "The Covered Bridge," Vermont Books, Middlebury, Vt., 1959, 150 pp. with photographs.

Dana, Richard T., "The Bridge at Windsor, Vt., and its Economic Implications," New York, 1926, 78 pp., illustrated.

DeMare, Eric, "The Bridges of Britain," London, 1954, 226 pp., illustrated.

Duggan, George, "Specimens of the Stone, Iron and Wood Bridges . . . of the United States Railroads," New York, 1850.

Edwards, Llewellyn N., "A Record of History and Evolution of Early American Bridges", University of Maine Press, Orono, Maine, 1959, 236 pp., illustrated.

Finch, James K., "A Century of American Civil Engineering," Transactions of the American Society of Civil Engineers, Vol. CT. 1953.

Flower, Frank Abial, "Reminiscences of General Herman Haupt," New York, 1901, 326 pp., illustrated.

Fugl-Meyer, H., "Chinese Bridges," London, 1937, pp.

Gies, Joseph, "Bridges and Men," Doubleday, Garden City, New York, 1963, 343 pp., illustrated.

Harmer, Henry W., "Covered Bridges of Harrison County, West Virginia," Charleston, 1956, 146 pp., illustrated.
Haupt, Herman, "The General Theory of Bridge Construction," New York, 1873, 268 pp.
Hopkins, H.J., "A Span of Bridges — An Illustrated History," David and Charles Publishers, Deven, England, Praeger Publishers, New York, 1970, 288 pp., illustrated.
Horst, Mel and Smith, Elmer, "Covered Bridges of Pennsylvania Dutchland," Lebanon, Pa., 1970, 42 pp., illustrated.
Hosking, William; Hann, James, and Weale, John and others, "The Theory, Practice and Architecture of Bridges," London, 1839.
Jacobs, David and Neville, Anthony E., "Bridges, Canals and Tunnels," Von Nostrand, 1968, 159 pp.,
Jakeman, Adelbert M., "Old Covered Bridges," Stephen Days Press, Brattleboro, Vt., 1935, 107 pp., illustrated.
Kenyon, Thedia Cox, "New Hampshire's Covered Bridges," Sanbornville, N.H., 1957, 140 pp., illustrated.
Kirby, Richard S., and Laurson, Philip G., "The Early Years of Modern Civil Engineering," New Haven, 1932, 324 pp., illustrated.
Lane, Oscar F. (edited by), "World Guide to Covered Bridges," National Society for Preservation of Covered Bridges, South Peabody, Mass., revised 1972.
Ketcham, Bryan E., "Covered Bridges on the Byways of Indiana," Lockland, Ohio, 1949, 226 pp., illustrated.
Mock, Elizabeth B., "The Architecture of Bridges," Museum of Modern Art, New York, 1949, 127 pp., illustrated.
Mohringer, Karl, "Bridges of the Rhine: Roman, Medieval and Modern," Messkirch, Baden, Germany, 1931, illustrated.
Morley, S. Griswold, "The Covered Bridges of California," University of California, Berkeley, 1938, 92 pp., illustrated.
Morse, Victor, "The Story of Covered Bridges in Windham County, Vermont," Brattleboro, Vt., 1937, 40 pp., illustrated.
Pancoast, Hazel and Chalmers, "Covered Bridges to Yesterdays," Newark, Ohio, 1959, 96 pp., photographs.
Peet, Creighton, "The First Book of Bridges," New York, 1953, 64 pp., illustrated.
Pope, Thomas, "A Treatise on Bridge Architecture," New York, 1811, 288 pp., illustrated.
Shank, William, "Historic Bridges of Pennsylvania, York, Pa., 1966, 56 pp., illustrated.
Spargo, John, "Covered Wooden Bridges of Bennington County," Bennington, Vt., 1953, 21 pp., illustrated.
Sloane, Eric, "American Barns and Covered Bridges," Funk and Wagnalls, New York, 1954, 110 pp., illustrated.
Sloane, Eric, "Our Vanishing Landscape," Funk and Wagnalls, New York, 1955, 107 pp., illustrated.
Smith, Hubert S., "The World's Great Bridges," Phoenix House, London: Harper, New York, revised 1964., 250 pp., illustrated.
Steinman, David B. and Watson, Sarah Ruth, "Bridges and Their Builders," Dover Publications, New York, 1957, 401 pp., illustrated.
Tyrrell, Henry G., "History of Bridge Engineering," (Published by the author), Chicago, 1911, 478 pp., illustrated.
Unger, Frederick F., "Old Bridges of Franklin County," Mercersburg, Pennsylvania, 1941, 98 pp., illustrated.
Wagemann, Clara E., "Covered Bridges of New England," Rutland, 1952, 152 pp., illustrated.
Walker, C. Ernest, "Covered Bridge Ramblings in New England," Contoocook, N.H., 1959, 114pp., illustrated.

Watson, Wilbur J., "Bridge Architecture," W. Helburn, Inc., New York, 1927.
Watson, Wilbur J. and Sara Ruth, "Bridges in History & Legend," Jansen, Cleveland, Ohio, 1937, 228 pp.
Wells, Rosalie, "Covered Bridges in America," New York, 1931, 135 pp., illustrated.
Whipple, Squire, "An Elementary & Practical Treatise on Bridge Building," New York, 1872.
White, W. Edward, "Covered Bridges of New Hampshire," Plymouth, 1942, 56 pp., illustrated.
Whitney, Charles S., "Bridges: A Study in Their Art, Science and Evolution," Rudge, New York City, 1929, 363 pp., illustrated.
Whittlesey, Charles W., "Crossing and Re-Crossing the Connecticut River," New Haven, 1938, 143 pp., illustrated.
Wood, De Volson, "Treatise on the Theory of the Construction of Bridges and Roofs," New York, 1873, 250 pp.

INDEX